SAFETY IN MICROBIOLOGY

THE SOCIETY FOR APPLIED BACTERIOLOGY
TECHNICAL SERIES NO. 6

SAFETY IN MICROBIOLOGY

Edited by

D. A. SHAPTON

H. J. Heinz Co. Ltd., Hayes Park, Hayes, Middlesex, England

AND

R. G. BOARD

Department of Biological Sciences, University of Bath, Claverton Down, Bath, England

1972

ACADEMIC PRESS · LONDON · NEW YORK

ACADEMIC PRESS INC. (LONDON) LTD
24/28 OVAL ROAD
LONDON, NW1

U.S. Edition published by
ACADEMIC PRESS INC.
111 FIFTH AVENUE,
NEW YORK, NEW YORK 10003

Library of Congress Catalog Card Number : 77–189937
ISBN : 0–12–638860–1

Printed in Great Britain by
Cox & Wyman Ltd., Fakenham, Norfolk, England

Contributors

V. G. Alder, *United Bristol Hospitals, Bristol BS2 8HW, England*

J. N. Andrews, *Centre for Nuclear Studies, University of Bath, Bath, BA2 7AY, Somerset, England*

D. K. Blackmore, *M.R.C. Laboratory Animals Centre, Carshalton, Surrey, England*

I. J. Bousfield, *National Collection of Industrial Bacteria, Torry Research Station, Ministry of Agriculture, Fisheries and Food, Aberdeen, Scotland*

R. C. Codner, *Microbiology Section, School of Biological Sciences, University of Bath, Bath BA2 7AY, Somerset, England*

T. Cross, *Postgraduate School of Studies in Biological Sciences, University of Bradford, Bradford, Yorkshire BD7 1DP, England*

W. A. Cox, *Unigate Central Laboratory, Western Avenue, Acton, London W.3, England*

F. A. Dark, *Microbiological Research Establishment, Porton Down, Salisbury, Wiltshire, England*

H. M. Darlow, *Microbiological Resaerch Establishment, Porton Down, Salisbury, Wiltshire, England*

R. Elsworth, *New Brunswick Scientific Co., Inc., New Brunswick, New Jersey 08903, U.S.A.**

C. G. T. Evans, *Microbiological Research Establishment, Porton Down, Salisbury, Wiltshire, England*

G. T. Harper, *Microbiological Research Establishment, Porton Down, Salisbury, Wiltshire, England*

R. Harris-Smith, *Microbiological Research Establishment, Porton Down, Salisbury, Wiltshire, England*

E. G. Harry, *Houghton Poultry Research Station, Houghton, Huntingdon, England*

A. C. Hill, *M.R.C. Laboratory Animals Centre, Carshalton, Surrey, England*

Margaret C. Hooper, *Quality Control Microbiology, The Boots Company Ltd., Nottingham, England*

* Address for correspondence: 25 Potters Way, Laverstock, Salisbury, Wiltshire, England.

D. J. Hornsey, *Centre for Nuclear Studies, University of Bath, Bath BA2 7AY, Somerset, England*

J. Lacey, *Rothamsted Experimental Station, Harpenden, Hertfordshire, England*

J. E. Lewis, *Unigate Foods Ltd., Bailey Gate, Sturminster Marshall, Wimborne, Dorset, England*

A. R. MacKenzie, *National Collection of Industrial Bacteria, Torry Research Station, Ministry of Agriculture, Fisheries and Food, Aberdeen, Scotland*

Isobel M. Maurer, *Disinfection Reference Laboratory, Central Public Health Laboratory, Colindale Avenue, London NW9 5HT, England*

Mary McClintock, *Ministry of Agriculture, Fisheries and Food, Agricultural Development and Advisory Service, Coley Park, Reading, England*

J. P. Mitchell, *United Bristol Hospitals, Bristol BS2 8HW, England*

J. Pepys, *Institute of Diseases of the Chest, Brompton, London S.W.3, England*

R. Smart, *Quality Control Microbiology, The Boots Company Ltd., Nottingham, England*

R. Spencer, *The British Food Manufacturing Industrial Research Association, Randalls Road, Leatherhead, Surrey, England**

D. F. Spooner, *Quality Control Microbiology, The Boots Company Ltd., Nottingham, England*

J. E. D. Stratton, *Microbiological Research Establishment, Porton Down, Salisbury, Wiltshire, England*

G. Sykes, *Quality Control Microbiology, The Boots Company Ltd., Nottingham, England*

J. M. Wood, *The British Food Manufacturing Industrial Research Association, Randalls Road, Leatherhead, Surrey, England*

M. J. Yalden, *W. H. S. (Pathfinder) Limited, Solent Road, Havant, Hampshire, England*

* Present address: J. Sainsbury Ltd., Stamford House, Stamford Street, Blackfriars, London, S.E.1., England.

Preface

THIS volume includes contributions to the Autumn Demonstration Meeting of the Society for Applied Bacteriology, held on 28th October, 1970 at the Department of Biology, Brunel University. It is Number 6 of the Technical Series and it continues the Society's policy of providing workers in a particular field with the opportunity firstly of demonstrating methods and techniques to Members and guests of the Society and secondly of describing these in a book which is intended for use at the bench. The Demonstration had as its central theme, Safety, and the organizers of the meeting adopted a liberal interpretation of this aspect of microbiology so that safe working with chemicals, radioisotopes, *etc.*, is considered along with those techniques which have evolved for routine handling and storing of saprophytic microorganisms and highly virulent pathogens. The liberal interpretation of safety is reflected in this book and it should be of interest not only to microbiologists of long standing but also to the many persons who find employment in microbiology after training in other disciplines where there is perhaps less requirement for the worker to be constantly aware of the risk of cross contamination or infection.

We wish to thank all the demonstrators for the great effort which they took both in the preparation of the exhibits and the chapters in this book.

Our particular thanks go to Professor J. D. Gillett, Mr. F. G. B. Jones, and Mrs. S. Bannerman and other members of the Biology Department of Brunel University for all their help with the laboratory arrangements for the Demonstration.

April, 1972

D. A. SHAPTON
R. G. BOARD

Contents

Safety in the Microbiological Laboratory: An Introduction

H. M. Darlow

Microbiological Research Establishment, Porton Down, Salisbury, Wiltshire, England

Since long before the days of Louis Pasteur, workers in the field of pathogenesis have infected themselves in an endeavour to establish those rules that have since crystallized as Koch's Postulates. John Hunter, for example, deliberately and successfully initiated in his own tissues that process, which we now associate with a positive Wassermann reaction, by means, which though relatively painless, were certainly not in the least romantic. It is unfortunate that with increase in knowledge and the availability of *para*-human models laboratory workers still persist in infecting themselves, and more so in that the process is now unintentional; to quote Chatigny and Clinger (1969): "It may be stated without fear of contradiction that every infectious microbial agent which has been studied in the laboratory has, at one time or another, caused infection of operators. In some instances, laboratory infections out-number natural infections and have been the only known human infections." Sulkin (1961) recorded 2348 cases (mainly in the U.S.A.) of presumed laboratory acquired infection with 107 deaths. This may represent little more than the tip of the iceberg, since there is a very natural reluctance to advertise the results of carelessness or ignorance; and many obscure, minor, or subclinical infections must pass undiagnosed, especially when they occur as secondary cases, or in persons not directly concerned in work at the laboratory bench or in the animal house. In yet other instances the etiological connection between a disease and the victim's work may go unrecognized; one wonders how many people died of *Herpes* B virus infections before its connection with a minor mucosal eruption in non-human primates was established. The possibilities are far from being exhausted. New entities, such as Vervet Monkey Disease, may suddenly emerge; old ones, such as Serum Hepatitis, may assume a new significance with changing techniques; and the possible long-term hazards of handling such agents as the oncogenic and slow viruses is now dawning on the scientific conscience. These few examples alone have recently increased awareness of the need for laboratory safety, about which much has been written, but still relatively little read. This need to be interested should not be a mere function of the anti-pollution movement with its threat of a "Silent Spring", for it is based on the hard fact of morbidity figures which are, in effect, comparable with those of road accidents. Fortunately, there are remedies, albeit often ignored, which can save us from wasteful martyrdom to biological science, risk of justifiable public wrath at potentially dangerous ineptitude, and the "Silent Laboratory".

Pathogens and Non-pathogens

It can be argued that the first step in selecting safe procedures lies in the direction of determining what organisms are pathogenic, and what is the relative infectivity of these for man. Unfortunately, the problem is not as simple as this, as the factors involved are so multitudinous that it could well be simpler to regard **all** microorganisms as presenting some degree of hazard in one way or another. In actual practice, economic considerations all too often dictate the need for, and extent of, precautionary measures, until the occurrence of an expensive failure; though the reverse situation in which work is rendered so intolerably complicated by precautions, often devised from the depths of an armchair, that the worker is tempted to take short cuts, is by no means unknown. A happy medium can be struck only after due assessment of all available considerations, which fall into three main categories, the nature of the organism, the ecosystem of the laboratory (health, hygiene and design) and the mechanics of the experimental techniques employed therein. Let us consider, firstly the organisms. These fall into 7 natural groups.

Established human pathogens

Agents in this category cause conditions ranging from rapidly fatal disease to minor indispositions, or purely localized and self-limiting lesions; they may, or may not, be transmissible to other human contacts, either directly, by vectors, or on fomites. The risk involved in handling them will depend not only on this, but also on viability, virulence, infectivity, portal of entry, size of challenge, the immune status of individuals or populations at risk, and factors specific to laboratory conditions, such as hygiene and experimental procedures; and also on the possession of effective, specific therapeutic measures should the worst occur. The implications of some of these factors will become apparent later, but for the present it must suffice merely to state the obvious—that all established human pathogens should always be handled with circumspection, and preferably with equal caution.

"Avirulent" strains of pathogens

The avirulence of a strain is largely a matter of degree, and is related again to the dose and portal of entry, and to the resistance of individual hosts and host species. A classic example of the latter is that of Strain 19 of *Brucella abortus*, employed as a live vaccine in animals, but which produces florid undulant fever in man. In addition, such events as the accidental substitution of a virulent for an avirulent strain, or enhancement of virulence by mutation or phage transduction must not be ignored. Here again are good reasons for caution.

Pathogens of animals

A very high proportion of pathogens in this category are transmissible to man, some so disastrously that one is inclined to forget that man is not the primary host (e.g. Plague, Tularaemia, Bovine Tuberculosis, and possibly Yellow Fever); whilst of the remainder some are sufficiently horrific (e.g. Rabies, *Herpes* B and Vervet Monkey Disease) to demand extraordinary precautions. Once again there is no excuse for relaxation, and it must not be forgotten that man can, and frequently does, act as an active or mechanical vector of pathogens in animal husbandry, apiary, sericulture, menageries, laboratory animal houses, veterinary practice, and even fisheries.

Plant pathogens

James Thurber stated that his great uncle, Zenas, died in 1866 of the disease that was killing the chestnut trees. Though this claim is unique, not to say dubious, there is no doubt that man is a sufficiently effective mechanical vector of plant pathogens to have stimulated the Ministry of Agriculture to forbid the laboratory handling of a long list of organisms except under licence (Anon, 1965*a*, *b*), issued subject to the provision of satisfactory safeguards. Septicaemia due to *Erwinia* has been reported (Mildvan *etal.*, 1971).

Facultative pathogens

To what extent some strains of *Proteus, Klebsiella, Aerobacter, Escherichia, Paracolobactrum* and *Pseudomonas* are actually primarily pathogenic is still debatable. There can be no doubt, however, that massive and fatal infection can occur in individuals whose normal response has been altered by disease, trauma, irradiation, immuno-suppressive drugs or antibiotics, though in other instances such synergic factors seem to be lacking. Drawing a hard and fast line between safe and unsafe strains in the absence of well established typing techniques (such as exist fairly comprehensively for Gram positive cocci for example, and, indeed for a few of the above mentioned organisms) is difficult, and one must take into account the source of the organism. Obviously a strain of *Bacillus cereus* or *Serratia marcescens* derived in pure culture from a case of meningitis at *post-mortem* cannot be treated as casually as indistinguishable wild isolates. Such admittedly rare events (and both are recorded) may easily be dismissed as "just one of those things", but could as easily represent an early stage in the evolution of a host/parasite relationship. It is significant, too, that this group of organisms as a whole tend to be poor antigens and resistant to antibiotics. Prophylaxis and therapy, therefore, provide poor defensive prospects.

Into this category, also, must fall organisms which, whilst not pathogenic in themselves, produce toxins under certain abnormal circumstances (e.g. *Clostridium* spp in anaerobic wounds, gut contents, meat pies, etc.), or which promote allergic responses (e.g. the micro-fungi of Farmer's Lung; see p. 151). Conditions of laboratory cultivation generally increase the pathogenic potential of these organisms, and often intentionally as in the production of toxins in toxoid manufacture.

Non-pathogens

Here belong all those organisms which have never started the hare of suspicion running. The list is very long and includes many organisms of major importance in human ecology, such as in sewage disposal, fermentation, nitrogen fixation and antibiotic production. One must be on guard for unexpected complications, but in the main their importance in the laboratory lies in their capacity, if carelessly handled, to create costly contamination problems, particularly as in many cases their relatively simple nutritional requirements and natural resistance to environmental hazards render them difficult to control.

Oncogenic viruses

The association between some viruses and tumour production in experimental and domestic animals is now well established, and, whilst progress in the human field is naturally not so well advanced and almost entirely confined to benign tumours, it is reasonable to expect a similar situation to exist. Malignant lymphomas have, in fact, occurred in laboratory workers involved in research with animal tumour viruses, and, whilst this is still regarded as coincidental, this and similar episodes have been sufficiently disquieting to have prompted the U.S. Department of Health to issue a code of practice on the safe handling of tumour viruses and cells (Anon,

1970). In the same document references are given to aerosol transmission of oncogenic viruses, excretion in urine and faeces of laboratory animals, and the detection of antibodies in the serum of laboratory staff. It is clear, therefore, that the laboratory ecology of at least some oncogenic viruses differs in no way from that of other types of infective agent.

A property of some viruses which may be relevant to safety is that of bimodal expression. Several viruses, including adenovirus strains of human origin, have been found to produce tumours in animals under certain laboratory conditions (e.g. in very high dosage in suckling mice). Spontaneous cases of this phenomenon are known in short-lived animals, but there is no clear evidence to suggest extension of the principle to humans. Nevertheless, it is probably fortunate that the organisms are already identified with established or "avirulent" pathogens. Until proved otherwise, therefore, it must be assumed that oncogenic animal viruses may constitute an oncogenic hazard to man, though the picture is complicated by the discovery that a virus infection can be expressed in different hosts by entirely different pathology.

The Mechanism of Infection

A more detailed attempt to categorize the contents of Bergey's Manual would only result in a process referred to by the Oxford English Dictionary as "floccinaucinihilipilification", and would be less meaningful. Nosological breakdown tables of human laboratory infections have appeared frequently in the literature which, judging from the absolute numbers of casualities, suggest that certain agents, notably *Francisella tularensis, Brucella* spp and *Coxiella burnetti*, are more infective than others. Whilst these assessments are probably close to the truth, they do not take into account variables such as the number of persons at risk, the relative hazards of the different handling techniques employed, and local hygiene or medical factors, and it may well be that more rarely handled agents are equally infective. The argument has, therefore, turned full circle to the original contention that most, if not all, microorganisms commonly handled in laboratories should be treated with respect for reasons ranging from their lethal potential to their nuisance value. Nevertheless, some ground has been cleared, and it can at least be said that the maximum hazard lies in handling potential killers and material of unknown potential sent to diagnostic laboratories (Sulkin *et al.*, 1963) for investigation.

The next logical step in assessing risk and how to reduce it is to consider transmission. This is divided into two stages: (a) dissemination, (b) invasion, or cause and effect; and as it is the latter that provides evidence that the former has occurred, it is to this that attention must first be directed, particularly in respect to human laboratory populations. Under natural conditions pathogens enter the human body through discontinuities in the skin, including all kinds of wound, insect bites, burns, blisters and dermatological conditions, by direct invasion of the mucosal surfaces of the gastrointestinal tract (i.e. by ingestion), mouth, nasopharynx, eyes and lower urogenital tract. Only in exceptional circumstances does the lower respiratory tract constitute a primary portal of entry, most pulmonary infections being secondary to systemic infection or downward extension from the throat. Of the exceptions the three classic examples are pulmonary tuberculosis, pneumonic plague and inhalation anthrax, in which the predisposing common factor is overcrowding in insanitary, ill-ventilated spaces, where the risk of inhaling infective material is greatly increased (see p. 167). This factor is also common to many laboratories, where all three examples have in fact taken their toll.

It has been found repeatedly that only a minority of recorded human laboratory infections are preceded by an overt accident, of which few have involved aerosol dissemination; the remainder of such cases followed accidental hypodermic inoculation or ingestion during mouth pipetting, and similar episodes of which a worker is conscious. Nevertheless, it has been concluded by Sulkin (1961) and other specialists in this field that in cases unassociated with an accident (approximately 80%) the prelude has been unsuspected aerosol generation during ordinary laboratory procedures. This does not imply pulmonary or even pharyngeal pathology, as in the case of plague, etc., since the tissue preference of many pathogens does not necessarily involve this, but animal and both accidental and intentional human challenge have adequately demonstrated that the great majority of pathogens can gain access by the pulmonary, or at least respiratory, portal of entry regardless of their natural mode of transmission. The arboviruses (Hanson *et al.*, 1967) provide an outstanding example of this.

Almost all common laboratory bench techniques and accidents produce aerosols in varying degrees, and the literature on the subject is, in fact, voluminous. References to it are given by Kenny and Sabel (1968) who confirmed previous findings and determined the concentrations and particle size ranges of aerosols generated during common procedures and simulated accidents. Infective laboratory aerosols can be divided more or less neatly into three categories.

Droplet nuclei

When disruptive stresses are applied at the surface of a liquid, droplets are detached. In laboratory air these evaporate extremely quickly (Green and

Lane, 1964), leaving nuclei of suspended and dissolved solids. The smaller the solid content of the original liquid, then the smaller will be the resultant nuclei for any given droplet size. The smaller the nuclei, the longer will they remain airborn (Green and Lane, 1964) and the greater their chances of dissemination and of being inhaled. Furthermore, the smaller the particle, the deeper into the respiratory tract can it penetrate (see p. 167). Much has been published in the last quarter of a century on this latter aspect which demonstrates that at least some agents can infect in smaller dosage and with more drastic results and shorter incubation periods, when the particle diameter is less than 5 μ and pulmonary penetration achieved, than when larger and retained in the nose (see p. 167). This suggests that pipetting serial dilutions of bacterial suspensions and subsequent plating out, when accompanied by careless splashing and frothing, is particularly hazardous. Casualty figures tend to confirm this, though the manipulation of relatively viscous slurries (e.g. pus, blood, sputum, infected tissue cultures, etc.) is by no means free from risk (Kenny and Sabel, 1968).

Dried materials

The disintegration of lyophilized cultures, crusts on stoppers and screw-caps, dry colonies and exudates, and the dehiscence of fungal sporangia are all potent sources of small, light particles with an unusually concentrated content of infective units. It can also be shown that the act of cutting deep-frozen tissue sections liberates minute fragments of material that are widely disseminated and may also present a bio-hazard.

Vector dusts

These consist of fur, down, skin scales, cage litter, "fluff" (e.g. cotton wool plugs and hospital blankets) and heterogeneous particulate detritus usually dignified as house dust. The particles are relatively large, but buoyant. Their presence is readily observed, and accumulations are surely a sign of bad housekeeping. They may be contaminated either at source or by subsequent fall-out of smaller particles. Whilst they may well constitute a lesser respiratory hazard than the foregoing groups in some respects, because of the ease with which they are filtered out in the nasal passages, there is little doubt that they can contribute to the transmission of disease as exemplified by cage and hospital cross-infection and in the spread of certain epizootic diseases.

The infective potential of these three classes of particle depends not only on the factors already mentioned, but also upon the viability of the contained organisms. Apart from physical influences such as ambient temperature, humidity or light intensity, coupled with time, this will be affected by the co-existence of protective materials within the particle and indeed on the survival capacity of the organism itself. Thus, whilst an aerosol created by the atomization of an aqueous suspension of a labile organism may only constitute a transient hazard to persons in the immediate neighbourhood, resistant organisms and those protected by a film of dried serum, for example, may be equally hazardous to individuals separated both in time and space from the original emission, and dangerous accumulations can build up if the process is repeated. A little altruism is, therefore, not out of place.

Laboratory Techniques

Our forebears were well aware that air played a part in the transmission of disease, and coined the word "malaria" to describe the phenomenon; and though mosquitoes are now known to have been the culprits in the case of the Pontine miasmas, "drains" (or more often the lack thereof) have continued to evoke a justifiable degree of suspicion ever since. The basis of the relationship had to await the discovery of pathogenic microorganisms and means of detecting them in the air by sampling techniques, which are reviewed in this volume by Dark and Harper (p. 37). The actual processes whereby microbial cultures are aerosolized, and the laboratory techniques which predispose to this are set forth in great detail by Chatigny (1961), Darlow (1967*a*), and Chatigny and Clinger (1969). Present space does not permit a full review of the literature, and the reader should, therefore, refer to these works for further information, but the following list of aerosolizing processes is offered as a guide to determining which laboratory procedures can be rendered relatively safe by careful technique and which call for extraordinary measures.

The bursting of a liquid film

The disintegration of films of liquid in bubbles, froth, across bottle mouths and pipette tips, and in platinum loops is a potent source of droplet production, which is enhanced by an air pressure differential across the film, as in the initial stages of the bursting of a bubble or the parachute film developed (Fig. 1) when a drop falls from a height and air penetrating the film exerts a shearing action at the point of escape (Green and Lane, 1964). Most of the commonest laboratory procedures involve film-bursting in one form or another, and it is possibly the commonest cause of laboratory infections.

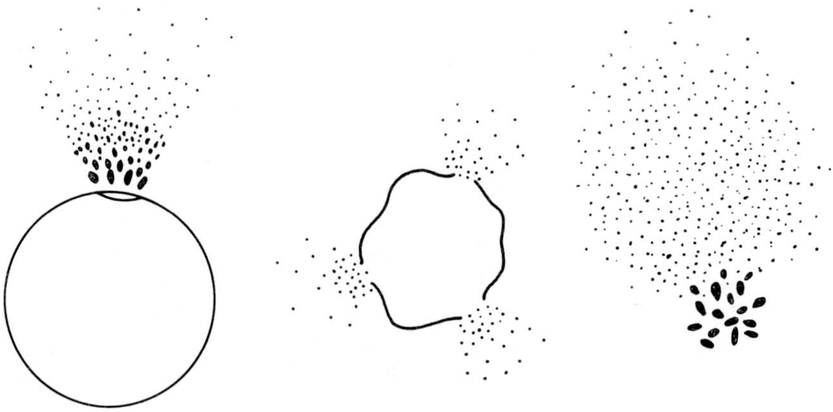

FIG. 1. Aerosol production on collapse of a bubble.

Careful handling can substantially reduce bubble-blowing to the accidental level, but the accidental level is perilously high.

The mixing of gas and liquid

The mild, purposeful and unavoidable mixing of liquid and gas as in shake cultures and fermenters, or merely the manual shaking of a test tube, leads to frothing with its attendant hazards. When more energy is put into such a system, however, bubbling plays a proportionately lesser role than shearing, resulting in direct expulsion of droplets into the atmosphere. The forceful ejection of pipette or syringe contents, which may already be charged with air bubbles, is a well known example of this process, which culminates in spray devices specifically designed for aerosol production. Forceful ejection of gas into liquid, *vice versa*, or of both into either presents a major hazard.

Vibration

The twanging of a platinum loop or hypodermic needle is as effective a catapult of microorganisms as a schoolboy's ruler with an ink pellet, and the principle has been intentionally employed in droplet production. By the same token, any other vibrating system presents some degree of hazard, and whilst it may be possible to avoid the above mentioned examples by the exercise of care and attention, appliances such as "ultrasonicators" require the use of safety appliances.

Falling drops

Apart from any disintegration (Fig. 1) that may occur during the fall of a droplet (Fuchs, 1964; Green and Lane, 1964), aerosol is produced more or less explosively on impact with a surface. The effect will be enhanced by gravitational acceleration or any other energy input. A drop falling naturally from a height of a few inches on to a smooth dry surface tends to spread out like a pancake with little tendency to rebound, but impact with a shallow liquid film, as on a wet agar plate or bench top for example, produces a coronet-shaped upward surge of droplets (Fig. 2) at the point where the

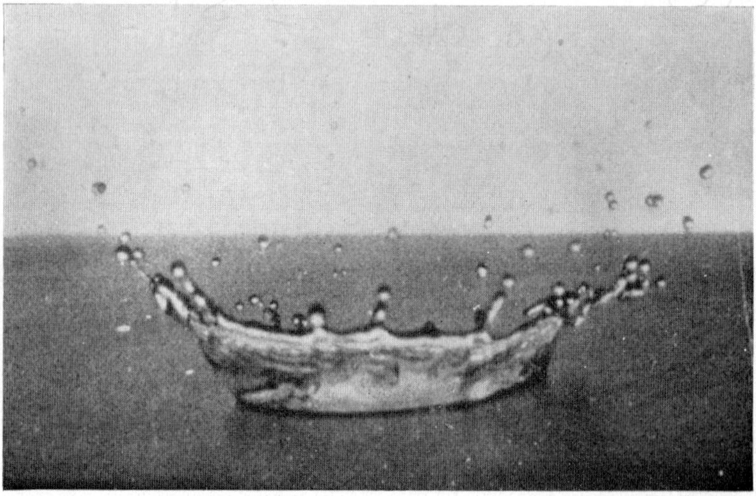

FIG. 2. Coronet produced by the impaction of a drop falling from a pipette tip on to a thin film of liquid on a smooth surface.

spread of the drop is arrested by the resistance of the film in very much the same fashion as a sea wave bursts against a breakwater. This is less evident in the case of impact with deeper liquid layers, where rebound takes the form of a central fountain or Rayleigh jet. In both instances satellite droplets large enough to fall back can repeat the process. Fall on to an **absorbent** surface, even though moist, is relatively safe; in fact open bench work **should be** carried out over a sheet of lint or similar material, moistened, but not soaked, with a germicide, if required. Here again much can be achieved by the exercise of care, though certain appliances such as fraction collectors, do demand more elaborate measures.

The "string of beads"

When two moist surfaces are separated, filaments of liquid are drawn off, which break up into droplets (Fig. 3). This process is common to many

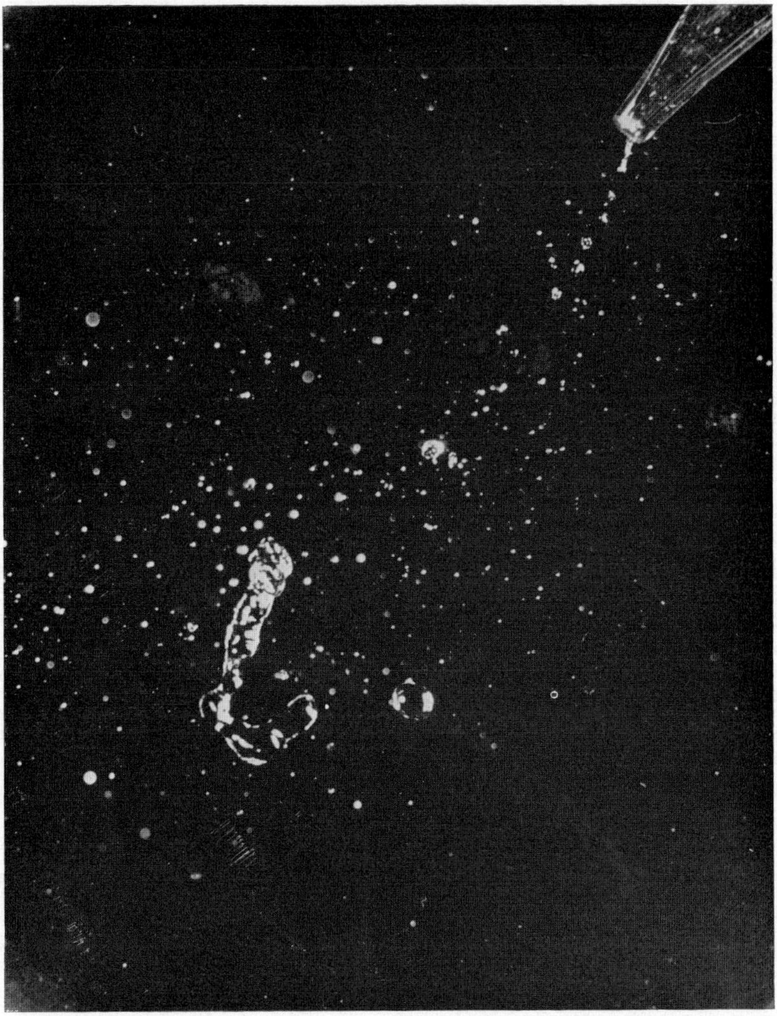

FIG. 3. Aerosol resulting from a bubble bursting at a pipette tip when the last drop was expelled. A "string of beads" is also visible emanating from the tip.

activities, such as slide agglutination techniques, or the simple withdrawal of a stopper from a tube and the plunger from a syringe barrel. It is often difficult to avoid, though in the case of stoppers and plungers, or the withdrawal of a hypodermic needle from a vein or a vaccine bottle cap, at least the operation can be carried out under the cover of an absorbent swab. The dropping and smashing of glass agar plate cultures (Fig. 4) is also a potent

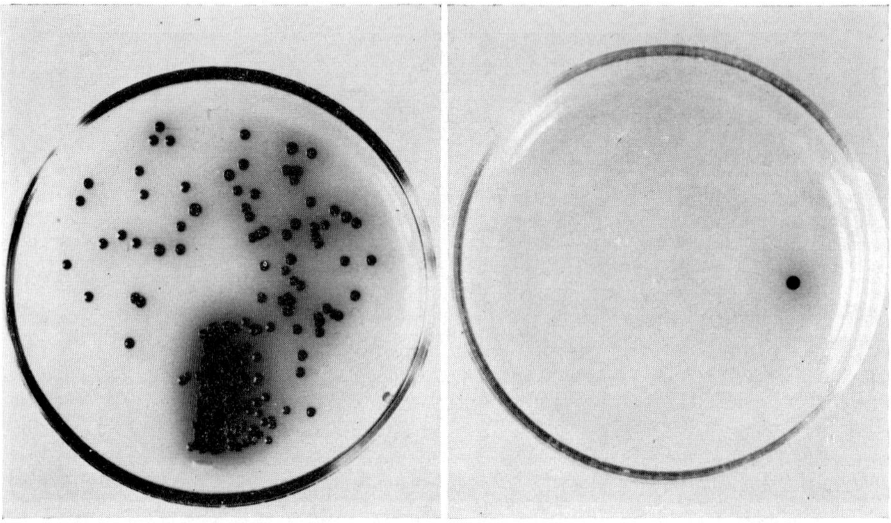

FIG. 4. Generation of an aerosol by dropping Petri dishes. A pile of 10 glass Petri dishes, each containing *c.* 50 colonies of *Serratia marcescens*, was dropped 3 ft on to a linoleum covered floor. A slit sampler positioned alongside the point of impact was used to monitor (1 ft^3/min) the aerosol (left hand plate). The right hand plate was used to monitor the aerosol produced when 20 plastic Petri dishes were dropped. (Reproduced by kind permission of the Editor, *Lab. Pract.*)

source of aerosol, and the mechanism involved is probably similar (Darlow, 1960; Barbeito, Alg and Wedum, 1961).

Centrifugal force

Centrifugal force is used intentionally to produce and disseminate liquid droplets in many contexts, ranging from crop spraying and laboratory aerosol studies to the schoolboy trick of throwing an egg into an electric fan; yet the principle is all too often ignored in cases where centrifugal separation, or homogenization, etc., are the primary aims.

All types of laboratory centrifuge, particularly angle-head rotors, can generate aerosols. Even hermetically sealed rotors cannot be relied upon not to leak at high speeds under vacuum, though good design and maintenance (in the sense of servicing) of the seal can reduce the hazard considerably. Where toxic materials are concerned, such rotors and sealed buckets should only be opened in a safety cabinet, and an air filter should be incorporated between the bowl and vacuum pump to prevent contamination of the pump oil and redispersal of the agent in oil mist. All other types of centrifuge can be regarded as potentially dangerous and should be operated in ventilated cabinets (see p. 30), or in ventilated rooms provided with means of respiratory protection for the operator and adequate means of decontamination. Homogenizers can be rendered as leak-proof as high-speed rotors, but some types are less efficient and all have to be opened up ultimately. Again a cabinet system is essential.

"Sizzling"

Sizzling is produced whenever a platinum loop or other inoculating tool is flamed or plunged hot into broth. Viable material is ejected both as visible

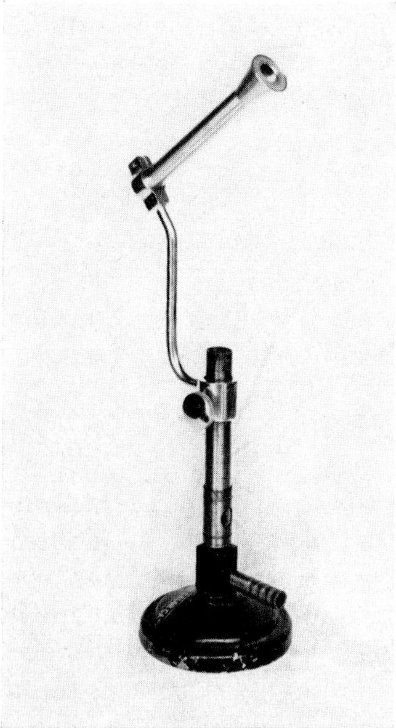

FIG. 5. A device for flaming platinum loop. A stainless steel tube (length, 4·5 in.); bore, 7/16 in.; wall thickness, 1/32 in.) has one end expanded to form a funnel which is partially occluded by an anular baffle—the latter acts as a spark guard. The tube is held at an angle of 45° in the hottest part of the flame. (Reproduced by kind permission of the Editor, *Lancet*.)

globules that fall back on to the bench top or hands of the operator, but a measurable aerosol is also produced. Cool loops should always be to hand, and, whilst various types of Bunsen hood (Fig. 5) have been described (Darlow, 1959), the flaming hazard is best eliminated by the use of a safety cabinet (see p. 25).

Electrostatic effects

This is a somewhat more speculative topic inserted to stimulate thought. Electrostatic repulsion has, in fact, been utilized as a means of experimental aerosol generation, and could play a part in maintaining them; one wonders what function it may play, in the explosive dissemination of fungal spores. Attraction, on the other hand, is an important factor in several types of air sampler (see p. 44) and in the efficiency of air filter media, albeit in some measure neutralized by high humidity and ionizing radiation. Charged plastic objects, however, can be shown to be attractive to aerosol particles, as evidenced by the use of plastic Petri dishes in an Andersen sampler for example, and could, therefore, present a more potent tactile hazard than uncharged surfaces. The very fact that electrostatic effects are employed to disseminate and capture organisms automatically implies hazards, even though obvious examples have yet to be demonstrated.

Gross splashes

The hazard here is obvious and involves all the foregoing principles plus ingestion, inoculation (in the sense of "into the eye") and bad housekeeping. It is, nevertheless, necessary to mention it, not only for the sake of completeness, but also to lead into the next phase of the argument.

Physical Protection

It is not on record, but Confucius, or one of his latter day compatriots, must surely have said at some time that there were three ways of avoiding being eaten by a tiger, namely: to wear a suit of armour; to go out and shoot it, or to keep it in a well-barred cage. The same law applies equally to biohazards, substituting more stilted terminology—protective clothing, decontamination and containment techniques. Of these, the third is by far the most important, since it largely eliminates the inconveniences and uncertainties of the other two; yet, strangely enough, it is the most often neglected, partly on account of the initial cost of equipment, and partly because of the mistaken idea that it restricts the freedom of the worker as much as the agent. In a few instances this is undoubtedly true, such as in the handling of large experimental animals, where the worker must fall back on more elaborate personal protection and decontamination measures than would otherwise be required, but, in the main, confining the agent is surely the most logical solution. It need not be inconvenient, and in the long run is not fraught with the doubts and complexities inherent in individual protection, which automatically imply a defensive, almost defeatist, outlook, and which is far more prone to the common human failings of lack of judgement or plain gambling with fate.

Undoubtedly, the most important piece of equipment is the **safety cabinet** (Barbeito and Taylor, 1968; Darlow, 1967*b* and Evans and Harris-Smith, see pp. 21–36). Unfortunately, this is a somewhat loose term embracing devices ranging from expensive and complicated hermetically sealed glove boxes, relying on negative pressure for safety, to simple hoods of the chemical fume cupboard type which, though relatively inexpensive and convenient, can present serious hazards. However, a happy medium can be struck provided that certain design criteria are adopted.

Whilst the shape (see pp. 29–32) is largely dictated by the nature of the work to be performed therein, the volume should be kept as small as convenient so as to ensure a high rate of air exchange, and hence, rapid elimination of aerosol, fumes and heat. All parts of the interior should be within arms reach of a port for preference to facilitate internal manœuvres, though a little extra depth is convenient for the accommodation of apparatus not in immediate use. Such can be mobilized with tongs or a *croupier*'s rake. It is better to employ a number of small cabinets, connected together by ventilated air locks, if desired, than one large one.

The arm ports should be restricted in size to ensure a **minimum** air flow through them of 100 linear ft/min. Very much higher velocities are required across open fronts of the fume cupboard type, if turbulence and localized retrograde flow are to be avoided. Open fronts also predispose to the escape of liquid spills. Double sided cabinets, though attractive in concept, can present a hazard, since they involve a multiplicity of ports and a risk of air being pushed right through the cabinet by the movements of an operator or draughts within the laboratory.

The exhaust fan or fans must be powerful enough to maintain the necessary ventilation rate and port velocities against the resistance of air sterilizing equipment, the cabinet itself, and, if the effluent air is to be discharged outside the building, any negative pressure produced by the laboratory ventilation system, or positive pressure due to external winds. Fans should be fully accessible and so fitted as to allow rapid replacement in the event of failure.

Though many methods have been, and indeed still are, employed to sterilize effluent air, the best all round method from all points of view (efficiency, energy requirements, emergencies, etc.) is undoubtedly high efficiency air filtration. Other methods have superficial attractions, and even value in limited contexts, but not here, and should be avoided, as anyone with an inquiring mind, scientific outlook and good test methods can soon confirm for himself. The filter should be mounted in an accessible position immediately on top of the cabinet so that its wall is virtually continuous with the cabinet walls and the walls of the effluent trunking or fan down stream. In this position it is fully exposed to fumigant gases and prevents contamination of exhaust fan and ducting. A coarse pre-filter should be fitted within the cabinet, where it can be safely changed whilst still contaminated, or removed to ensure penetration of the master filter by fumigants.

The only effective means of decontaminating the interior of a safety cabinet and its filter system *in situ* is by the use of gaseous fumigants, of which formaldehyde is both effective, safe, and simple to handle (for further details, see p. 33). It is essential therefore that the cabinet is capable of being rendered reasonably gas-tight when the ventilation is switched off. Other fumigants such as *Beta*-propiolactone, peracetic acid and even ethylene oxide have been used effectively, but have various disadvantages, and UV radiation, though frequently employed, is very unreliable for many and obvious reasons, and cannot penetrate the filter.

As the use of gauntlet gloves greatly enhances the safety factor, provision should be made for fitting them to the armports when high risk work is undertaken. This necessitates an air in-let suitably guarded by a valve or an in-let filter.

Electrical supplies, gas, compressed air lines, etc., should enter the cabinet *via* a small detachable panel on one wall, which can be drilled for additional services, if required, without modification of the main fabric, thus ensuring standardization and limiting the distribution of possible points of leakage. Lighting can with advantage be internal, provided moisture-proof terminals are fitted. Fluorescent strip lights are desirable because of low heat output and the fact that hot cathode UV lamps can be substituted and run off the same circuitry. Switch gear should be fitted with pilot lights, particularly when UV is installed, or when the exhaust fan is mounted remotely, where it cannot be heard.

Flow meters, pressure sensitive switches and other types of warning device may be fitted to the exhaust system, but the true acid test of ventilation efficiency is port-flow measurement with an anemometer. Apart from periodic bioassay of UV lighting, it is important to carry out filter efficiency tests at intervals, using a mono-dispersed aerosol of tracer organisms generated within the cabinet, air samples being taken through a sampling tube fitted conveniently between the filter and fan.

All these requirements may sound elaborate, but experience has shown them to be necessary for safety. There are of course many other points to be considered such as the resistance of constructional materials to disinfectants, solvents and other chemicals used in the cabinet, fire, gas explosions, electrical faults and UV radiation, but for these the reader must refer to the literature, or his own common sense.

Protective clothing—see also p. 92—is as prone to the whims of fashion as *feminene couture*, but the traditional theatre gown has stood the test of time and has much to recommend it. It can be removed single-handed, dirty side inwards, like skinning a rabbit; the cuffs are designed to tuck under surgical gloves and the bottom extends below the tops of gum boots into which spills and sharp instruments can otherwise fall. The fact that it does not adequately cover the back is of minor importance as compared with the hazards of removing a garment opening in any other way. Furthermore, as it is not an attractive garment, the worker will be prompted to remove it before leaving the laboratory. There is, of course, a case for more elaborate two-piece suits, special under-garments, caps, respiratory protection, and shower baths in specialized contexts, but where good containment techniques are used there seems little need for this in the majority of instances. Laboratory clothing should always be sterilized, preferably by autoclaving, before being unloaded on to the unsuspecting laundress; instances are on record where failure to do so has resulted in multiple infections.

Laboratory design is second in importance to containment techniques; indeed it is in itself a containment technique. Space does not permit a full review of the subject, which covers a very wide range of engineering and architectural disciplines, but Runkle and Phillips (1969) have recently compiled an invaluable account, phrased for the special indoctrination of the architect, thus filling a long felt want. Space only allows the inclusion of three useful guide lines here:

1. Provide elbow room.
2. Design-out dust-catching clutter and the need for maintenance staff to enter toxic areas.
3. Ensure a reasonable degree of air-tightness to confine aerosols and fumigants, and to avoid imbalance of the ventilation system, and ingress of insects and vermin.

It is also worth stressing here the dangers of overcrowding, which leads to lack of elbow room, too much apparatus and general clutter on the bench, the mixing of often incompatible disciplines, and the increase in number of personnel at risk in the event of hazardous procedures or accidents.

The subject of disinfection and sterilization has spawned a small library of books, of which the latest is edited by Hugo (1971), and is a valuable source of information on the kinetics and chemistry of germicidal methods.

It must suffice here merely to stress the point that many chemical germicidal agents have gaps in their activity spectra, which may constitute traps for the unwary. Their application in laboratory safety is largely covered in the quoted works on this topic, but where information is lacking, it is wisest to devote time to investigation rather than to take things for granted, or, regretably, to rely on over-enthusiastic sales talk. The annals of hospital infection are liberally scattered with instances of failure to do so. Chemical methods are frequently employed to disinfect contaminated glassware and disposables (see p. 53), but for many reasons this is unsatisfactory; their place should be as a prelude to, not a substitute for, autoclaving.

UV radiation, also, is still being used extensively in unsuitable contexts, and without adequate safeguards to ensure that it achieves its intended purpose. Many organisms are only slowly inactivated by a clean-walled lamp at full output, and are easily protected by films of debris, or shadows. Furthermore, the presence of visible spectral bands is no indication of germicidal activity, and hence regular cleaning and monitoring of installations is necessary. It must also be remembered that UV radiation is not only damaging to skin and eyes, but causes insidious deterioration of rubber and plastic components of apparatus (e.g. rubber gloves, electrical insulation and perspex windows) and can generate inhibitory substances in media.

Medical Aspects

Medical problems in the laboratory fall into 4 categories:

1. Supervision of general health and personal hygiene in respect of laboratory hazards. Are there contra-indications to exposure to infective hazards, such as corticosteroid therapy, skin conditions, open wounds, black-outs, or pregnancy; or has the worker some other trait liable to make him accident prone? Beards, for example, interfere with the efficiency of respirators and are known to carry a contamination risk, and long hair can readily catch fire or become entangled in moving mechanical parts.

2. Diagnosis, investigation, treatment or follow-up of known, or suspected laboratory infections. Here, it is also worthy of mention that periodic serological screening of apparently healthy staff can often be valuable in detecting otherwise unsuspected dissemination of antigenic agents as the result of unreported accidents or careless techniques—see also p. 172.

3. Emergency or anticipatory treatment, particularly in respect to potentially infective accidents.

4. Vaccination, which is placed last because it is by no means synonymous with immunization, unless accompanied by confirmatory serology.

It should not be omitted, however, because it is obviously politic to use an available vaccine, though its value may be uncertain, and it does not prevent the worker from becoming a passive carrier, or even an active, subclinical case.

The co-option of a medical adviser, therefore, is highly desirable. He should for preference be a qualified staff member familiar with laboratory work. Local public health authorities are generally most appreciative and co-operative, if kept informed of possible dangers, and their assistance is essential in the event of isolation requirements, and liaison with the hospital. The worker's general practitioner, too, should be made aware of the possibilities of laboratory infection in his patients, particularly in respect to more exotic conditions of which he may have little or no experience; and in at least one well known laboratory, workers carry a card warning the attendant doctor of the possible nature of an undiagnosed pneumonic or encephalitic condition, and giving the name and address of a knowledgeable referee. In addition it is of value to indoctrinate staff members themselves in the symptomatology of the conditions they may contract, not to mention interpretation of the abbreviations and code numbers applied to lethal agents, of the significance of which junior technicians are often alarmingly unaware. Training lectures with a medical flavour, therefore, are to be encouraged.

Conclusion

In conclusion it must be stressed that the foregoing philosophical and somewhat staccato diatribe has a firm basis in experimentation and, occasionally, bitter experience stretching over a good many years under circumstances common to research, production, and diagnostic laboratory work. There is an understandable tendency to dicotomize here, but, whilst the research worker may well be handling "hot stuff" in bulk, he at least knows, or should know, the nature of the hazard to which he is exposed, whilst the diagnostician may be quite unfamiliar with the risks attendant in handling the contents of that leaking sample tube, dead parrot, sack of bone meal or new intake of Vervet monkeys. Funds are generally more readily obtainable in research and production than in the diagnostic field, but there should not be one rule for the rich and one for the poor. The principles involved are the same, methodology basically similar and a life is a life.

References

ANON (1965*a*). *List of Cultures in the National Collection of Plant Pathogenic Bacteria, No. 6.* Harpenden: Plant Pathology Laboratory, Ministry of Agriculture.

ANON (1965*b*). *Destructive Pests and Diseases of Plants Order*. Article 4. London: H.M.S.O.

ANON (1970). *Biohazard control and containment in Oncogenic virus research*. Bethesda, Maryland, U.S.A.: U.S. Dept of Health, Education and Welfare, National Institutes of Health.

BARBEITO, M. S., ALG, R. L. & WEDUM, A. G. (1961). Infectious bacterial aerosol from dropped petri dish cultures. *Am. J. Med. Technol.*, **27**, 318.

BARBEITO, M. S. & TAYLOR, L. A. (1968). Containment of microbial aerosols in a microbiological safety cabinet. *Appl. Microbiol.* **16,** 1225.

CHATIGNY, M. A. (1961). Protection against infection in the microbiological laboratory. *Adv. appl. Microbiol.* **3,** 192.

CHATIGNY, M. A. & CLINGER, D. I. (1969). Contamination Control in Aerobiology. In *An Introduction to Experimental Aerobiology*. (Dimmick & Akers, eds), Chapter X (pp. 194–263). New York: Wiley.

DARLOW, H. M. (1959). A device for flaming platinum loops. *Lancet,* ii, 651.

DARLOW, H. M. (1960). Introduction to safety in the microbiological laboratory. *Lab. Pract.* **9,** 777.

DARLOW, H. M. (1967*a*). Safety in the microbiological laboratory. *Methods in Microbiology*; (J. R. Norris & D. W. Ribbons, eds). Vol. 1, pp. 168–204. London and New York: Academic Press.

DARLOW, H. M. (1967*b*). The design of microbiological safety cabinets. *Chemy. Ind.* 1914.

FUCHS, N. A. (1964). *The mechanics of aerosols* pp. 43–44. London: Pergamon.

GREEN, H. L. & LANE, W. R. (1964). *Particulate clouds, dusts, smokes and mists*. Chapter 3. (2nd Ed) London: Spon.

HANSON, R. P., SULKIN, S. E., BUESCHER, E. L., HAMMON, W. Mc.D., MCKINNEY, R. W. & WORK, T. H. (1967). Arbovirus infection in laboratory workers. *Science, N.Y.,* **158,** 1283.

HUGO, W. B. (1971). *Inhibition and destruction of the microbial cell*. London and New York: Academic Press.

KENNY, M. T. & SABEL, F. L. (1968). Particle size distribution of *Serratia marcescens* aerosols created during common laboratory techniques and simulated accidents. *Appl. Microbiol.* **16,** 1146.

MILDVAN, D., BOTTONE, E., HORSCHMAN, S. Z. & CORNELL, A. (1971). Septicaemia caused by microorganisms of genus *Erwinia*. *Mount Sinai J. Med.* **37,** 267.

RUNKLE, R. S. & PHILLIPS, G. B. (1969). *Microbial contamination control facilities*. New York: Van Nostrand, Reinhold.

SULKIN, S. E. (1961). Laboratory acquired infections. *Bact. Rev.* **25,** 203.

SULKIN, S. E., LONG, E. R., PIKE, R. M., SIGEL, M. M., SMITH, C. E. & WEDUM, A. G. (1963). Laboratory infections and accidents. *Diagnostic Procedures and Reagents*. (4th Ed.) p. 89. New York: Pub. Hlth. Assoc.

The Use of Safety Cabinets for the Prevention of Laboratory Acquired Infection

C. G. T. EVANS, R. HARRIS-SMITH
AND J. E. D. STRATTON

Microbiological Research Establishment, Porton Down, Salisbury, Wiltshire, England

Wherever infectious microorganisms or diseases are studied or treated, be it research laboratory, University or hospital, all persons associated with the building should understand that when an infectious microorganism or virus escapes, it may form an unknown, unseen airborne carrier of infection, fatal to the recipient (see p. 1). See for example the airborne transmission of smallpox in Meschede Hospital (Anon, 1970). Although the primary aim, to conduct research in complete freedom, is desirable, in microbiology it is prudent and indeed ethical for laboratory directors to ensure and provide against their staff unwittingly becoming part of their own experimentation. The ease with which infectious aerosols are generated by even the simplest laboratory techniques has been emphasized by Darlow (1969)—see also p. 6. Management should insist that microbiological disease hazards be contained at the source of any possible accidental leakage. Therefore it should be obligatory that a manager, officer or consultant with responsibility for safety should be appointed for all microbiological laboratories. It is also in the individual worker's own interest to avoid personal infection.

Although the labour codes of, for example, many States of the U.S.A., do place the onus of worker protection upon the employer, in the U.K. the laboratory worker is covered by neither the Factories Act nor the Shops and Offices Act. Only in the case of a few specified diseases such as tuberculosis, Anthrax, and Farmer's Lung, can compensation be claimed under the Industrial Injuries Act. Had the outbreak of Vervetosis (Marburg disease; Hennessen, Bonin and Mauler, 1968) occurred in this country, none of the 23 victims or their dependants could have claimed redress, except through the protracted and uncertain process of common law.

Bacteriological technique has evolved to protect the culture, not the worker, and most common laboratory techniques liberate aerosols which are potential hazards. Two general methods are used to contain aerosols: (a) the worker, in suitable protective clothing, is enclosed with the hazard, or (b) the worker is separated from the hazard which is enclosed in a safety cabinet or hood. The first method requires the worker to wear suitable protective clothing with a respirator or ventilated helmet and also requires the provision of air locks, etc., for his safe entry to and exit from the infected area (see Trexler, 1957). This method is neither practical nor necessary for most laboratory purposes and we shall only concern ourselves with the second method which involves the use of some form of safety cabinet.

There are many different types of the apparatus loosely referred to as "safety-boxes, cabinets, hoods", etc., not all of which are suitable or necessary for all purposes. It is, therefore, desirable to consider the following factors when choosing the appropriate type:

1. The infectivity, virulence and toxicity of the organism to be handled.
2. The degree of aerosol release likely to result from the envisaged technique.

The following additional points may have to be taken into account but should not override the decisions made to accommodate the previous factors.

3. Availability of suitable apparatus or the ease with which, if unsuitable, it may be modified.
4. The difficulty of the proposed operation in relation to the degree of protection required.
5. The construction and design of the laboratory.
6. The immune status of the experimenter and other staff.
7. The availability of effective treatment.

Reliance upon vaccines is not to be recommended; quite apart from variations of immune status from one individual to another, and in an individual, vaccines can only confer protection upon those vaccinated. Casual visitors and non-vaccinated staff remain vulnerable.

Efficiently designed microbiological safety cabinets interpose an effective barrier between the microorganism and the worker. There is a wide range of types available commercially, which vary considerably in design and efficiency and the same comments apply to the many types described in the literature. For convenience, we have separated the chief representative types into 3 classes:

1. Those designed primarily for the protection of cultures. (None of these are true safety cabinets.)
2. Those designed primarily for the protection of workers. (These are safety cabinets.)
3. Those designed for the protection of both cultures and workers. (These are aseptic safety cabinets.)

Protection of Cultures

Air not sterilized before discharge

No pressure differential.
Without ventilation

Hood. See e.g. Safety and vacuum enclosures, Tech. Bulletin A.G. 1970, S. Blickman Inc., 536 Gregory Ave, Weehawken N.J.

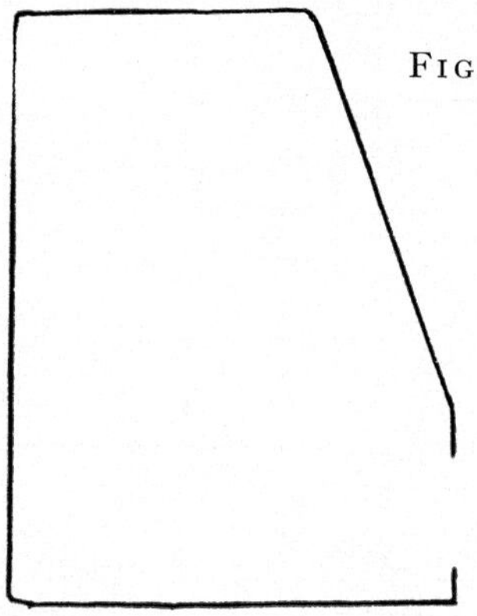

FIG. 1.

This type will only protect the worker's face against splashes and the culture against some extraneous dust particles. UV tubes can be fitted to control, partially, the bacterial contamination of the air; however, UV is not effective against spores, is absorbed by most surfaces except aluminium and some special paints, and the useful emission of the lamps is not necessarily correlated with the apparent intensity of illumination (Darlow, 1969).
Closed box. Movement of the gloves has a pumping action which can force aerosol in and out through leaks in the cabinet.

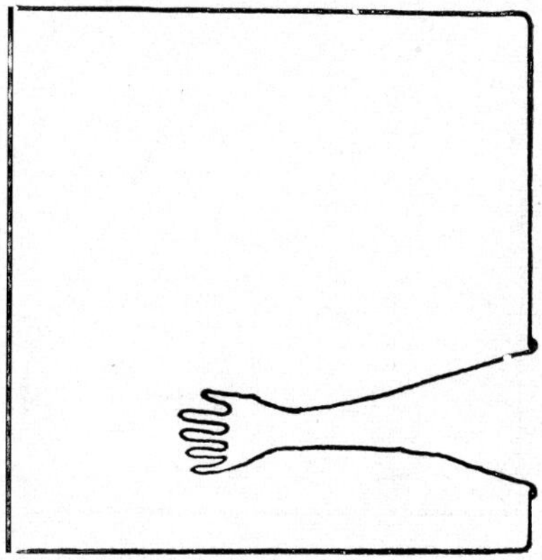

FIG. 2.

With ventilation

Open hood. Sterile air blown into hood (e.g. Harris-Smith, Pirt and Firman, 1963; Coriell and McGarrity, 1970).

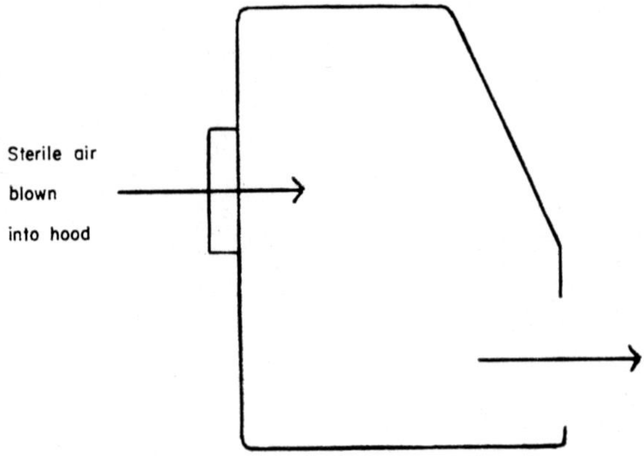

FIG. 3.

Although originally used in the electronics industry for dust free assembly of components, this type is now frequently used to protect cultures. Because any aerosol generated within the hood is blown at the worker, this hood is unsuitable for use with pathogens, allergens or toxins.

Closed box with air convected by burner. See for example Mackie and McCartney (1948).

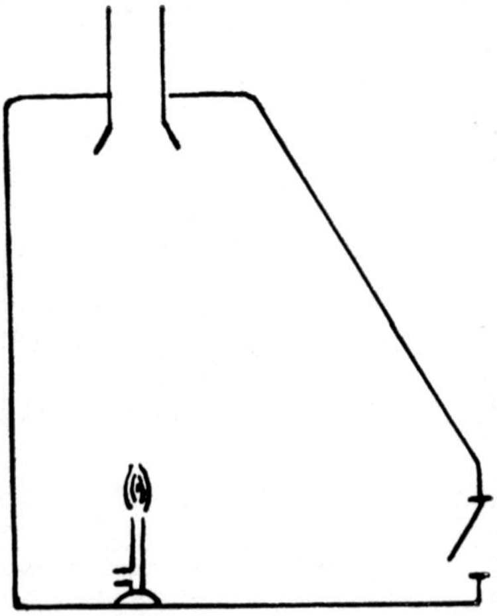

FIG 4.

Such a design which has not changed for more than 20 years, is illustrated by Cruickshank (1969) where it is recommended as a simple inoculation box and for preparation of blood plates, etc. "It is not recommended for dangerous pathogens such as the tubercle bacillus." As a general point, gas burners can be dangerous in confined spaces; we prefer to use spirit lamps where possible but if a gas burner is necessary, it should be used only in cabinets equipped with forced ventilation and should not be lit before the ventilation has been switched on. Gas burners should never be used in cabinets where air is recirculated.

Protection of Workers

Air drawn into hood and sterilized before discharge.
Open hood. See for example Williams and Lidwell (1957).

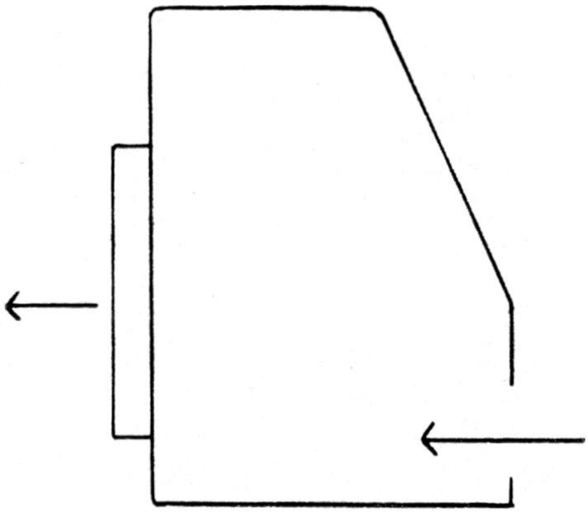

FIG. 5.

Hood with restricted access. See for example Barbieto and Taylor (1968).

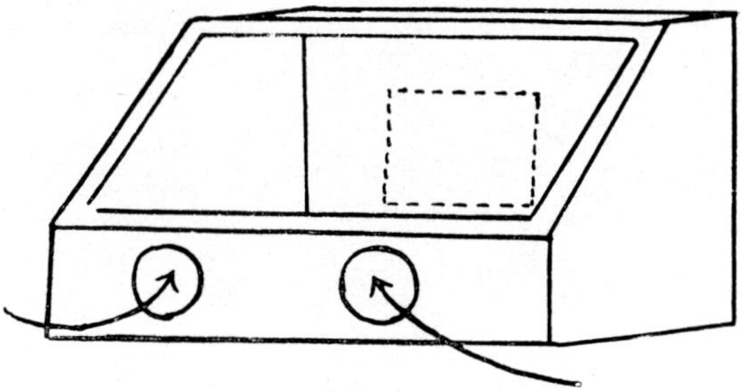

FIG. 6.

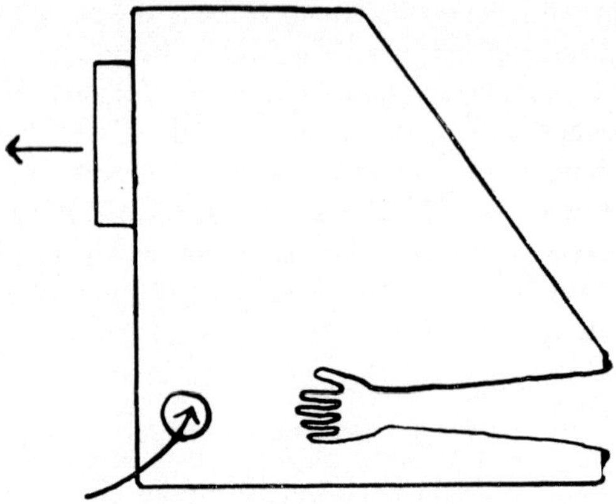

FIG. 7.

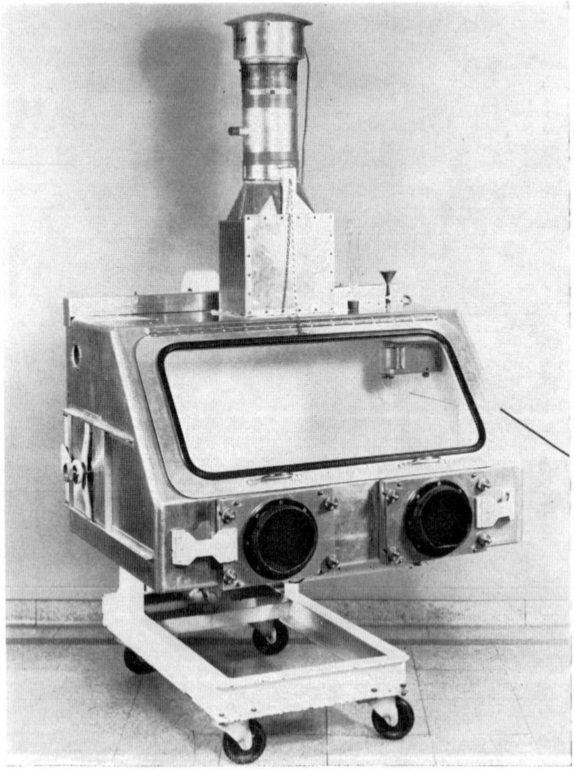

FIG. 8. The Porton safety cabinet. Air-flow rate, 100 CFM; air changes/h, 390, and cabinet clearance time, 10 min.

Neither of these types is recommended for containment of established pathogens, but if so used, the air velocity through any opening should be not less than 100 linear feet per minute to prevent the escape of aerosol when laboratory doors are used or when people walk past the front of the cabinet (Chatigny and Clinger, 1969). It should be noted that workers using hoods without gloves, should wear gowns which, having wrist bands, prevent aerosols being trapped up the sleeves (see Staat and Beakley, 1968). There is, of course, no containment of aerosol if there is a power or fan failure. As the culture is not protected, great difficulty may be found in trying to keep attenuated cultures of pathogens free from contamination.

Closed hood. Working access only through sealed gloves, air through separate inlet.

With this type, a pressure differential is maintained to ensure that any leaks are directed into the cabinet. An example (Fig. 8) of this type was described by Darlow (1967) and has now been used in MRE for more than 16 years. It gives excellent protection and has satisfied the conflicting demands of the staff.

Protection of Workers and Cultures

Open hood with curtain of sterile air. See Coriell and McGarrity (1968).

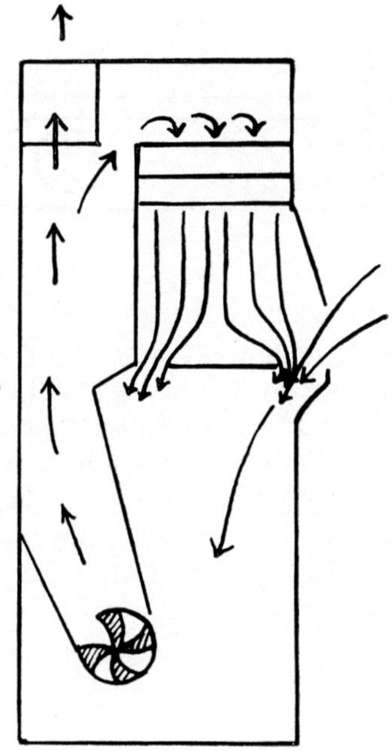

FIG. 9.

Although we have had no experience of this type of hood at MRE, there has been a number of reports in literature of their examination (e.g. Coriell and McGarrity, 1968; Staat and Beakley, 1968; McDade *et al.*, 1968). Their great attraction is the minimum of constraint to the operator; but there seems to be general agreement that the barrier afforded is not an adequate protection to worker or cultures, nor is it independent of the fan or power supply. They are probably adequate for handling non-infectious cultures and attenuated strains, but would not appear to be suitable for work with established pathogens where there is a possibility of aerosol generation.

Closed cabinet with access only through sealed gloves. Air sterilized as it enters and leaves the work area (e.g. Harris-Smith and Evans, 1968).

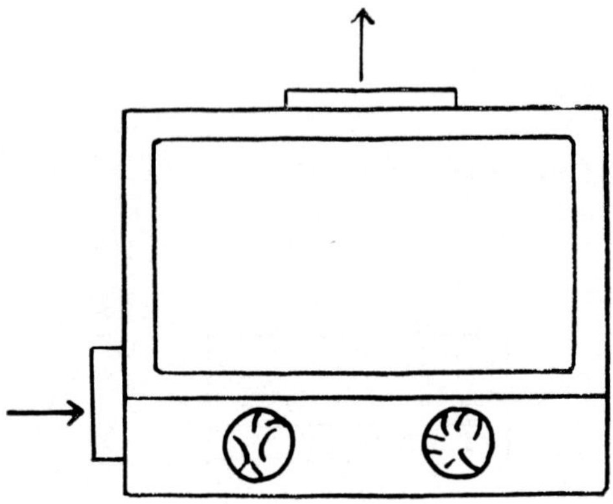

FIG. 10.

Provided that the construction and seals are leak proof, and that the filters or heaters used for sterilization are efficient, this type protects both worker and culture. Figure 11 shows a prototype made 7 years ago, but which proved to be so successful that it is still in use; from this have been derived the more specialized cabinets shown in Figs 12, 14, and 15 and described below. We shall hereafter refer to this type as an "aseptic safety cabinet".

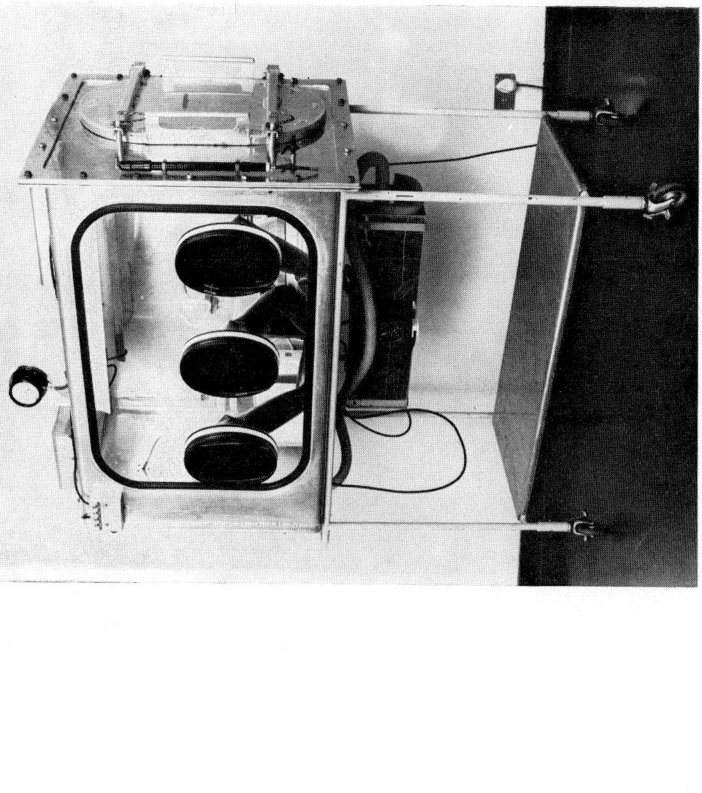

FIG. 11. Aseptic safety cabinet. Air-flow rate 56 CFM; air changes/h, 204, and cabinet clearance time, 25 min.

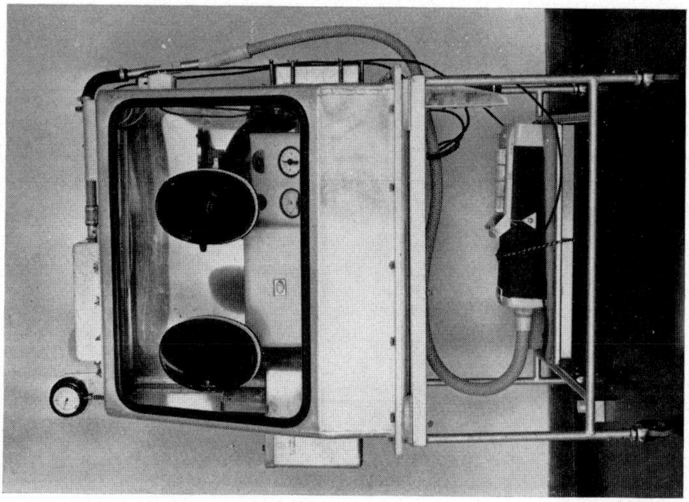

FIG. 12. Aseptic safety cabinet for M.S.E. Hi-Speed centrifuge. Air-flow rate, 67 CFM; air changes/h, 216; cabinet clearance time with centrifuge not running, 45 min; clearance time with centrifuge running, 30 min; this clearance time is less because the motor fan ventilates the apparatus voids.

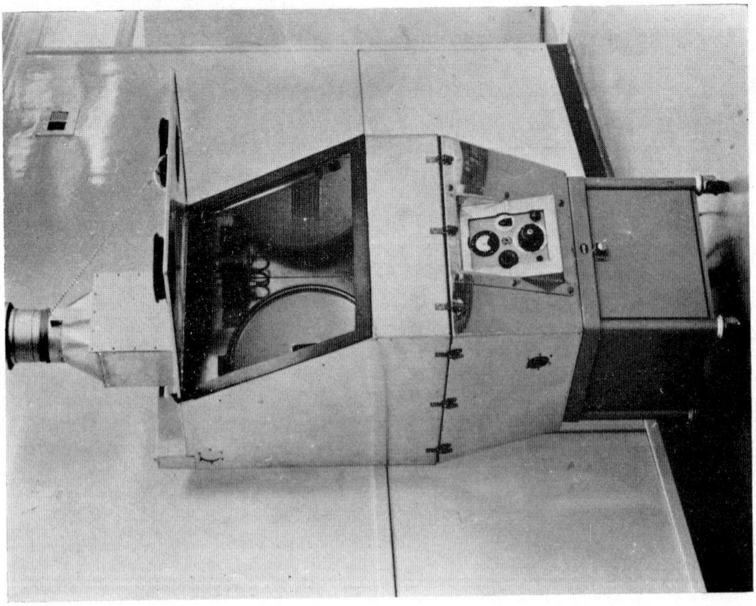

FIG. 13. Porton safety cabinet for M.S.E. 'Medium' centrifuge. Air-flow rate, 100 CFM; air changes/h, 390, and cabinet clearance time, 10 min.

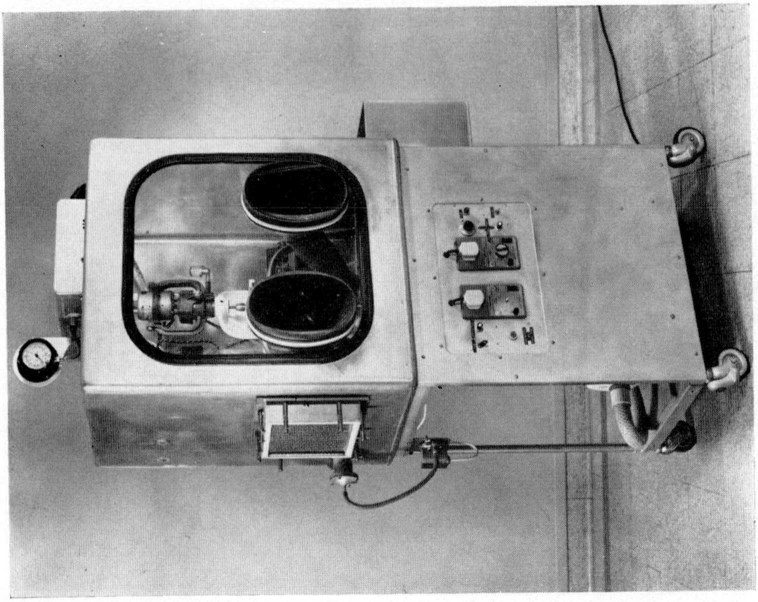

FIG. 14. Aseptic safety cabinet for continuous centrifuge. Air-flow rate, 75 CFM; air changes/h, 300, and cabinet clearance time, 40 min.

Sterilization of Ventilation Air

The two methods most used for sterilizing the ventilation air are either heat (e.g. Elsworth, Morris and East, 1961; see p. 61) or filters, (White and Smith, 1964), but UV radiation and electrostatic precipitators (Sykes, 1965) have occasionally been used. As we shall be considering these individual methods more fully in a separate publication, it is sufficient to say that, apart from their attractive simplicity, suitable filters provide adequate safety which is, of course, independent of power or fan failures. Filters must, like other forms of air sterilizers, be tested *in situ* under operating conditions with biological material similar in size and surface properties to that which they will be expected to retain in use (see Darlow, p. 1).

Ventilation Rate

The concentration of aerosol in the safety cabinet at any given moment will depend upon the rate at which it is being released and the rate and completeness with which the cabinet air is changed. A rapid rate of air change will quickly clear the cabinet unless there is a continuing source of aerosol release, as for example with a Sharples centrifuge. For such an apparatus, the ventilation rate is important and should be in excess of one air change per minute. However the completeness with which the cabinet air is changed is equally important if pockets of infected air are to be removed. Apparatus enclosed in a cabinet can itself create unventilated voids, as is well shown by the clearance times for the cabinet in Fig. 12. Such voids and unventilated spaces can often be demonstrated by the use of smoke tests (ammonium chloride is both convenient and effective). The clearance times for the various cabinets are given in the legends to the figures; they were obtained by deliberately generating an aerosol within the cabinet with the ventilation on and following its disappearance with a slit sampler (Bourdillon, Lidwell and Lovelock, 1948).

Transfer Methods

There are only two safe ways that articles may be admitted to, or removed from an infected cabinet. Either through a disinfectant liquid lock (dunk tank) or an integral autoclave. Otherwise the cabinet must be cleared by ventilation and disinfected before it is opened. Doors, openable fronts and glove ports can be used with discretion to admit articles if the air flow rates through the openings are sufficient, but air locks can be hazardous unless particular precautions are taken (see Darlow, Harris-Smith and Evans, in prep.). Suitable transfer canisters (Fig. 16) are used for passing materials through the dunk tank.

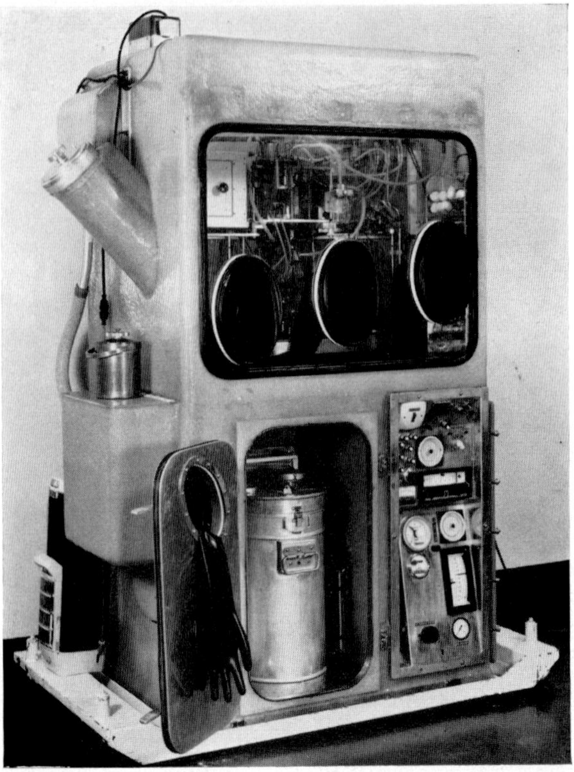

FIG. 15. The Porton mobile enclosed chemostat (POMEC). Air-flow rate, 87 CFM; air changes/h, 222 in process chamber, 834 in transfer chamber; process chamber clearance time, 10 min, and transfer chamber clearance time, 5 min.

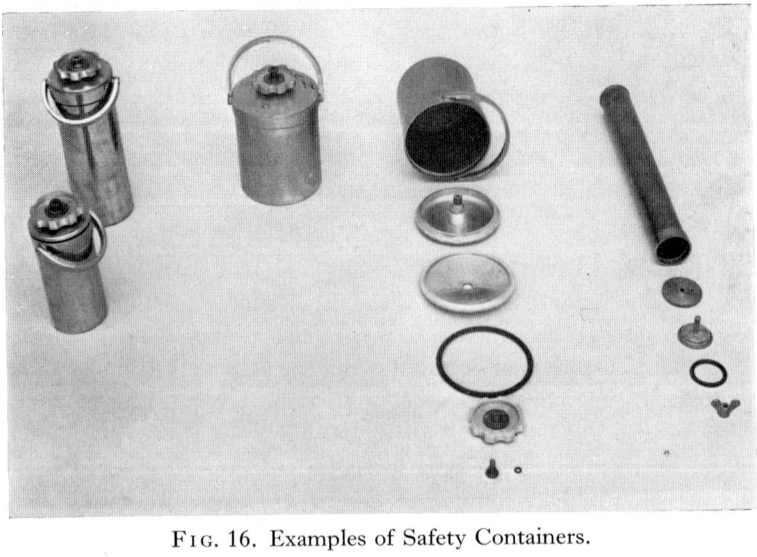

FIG. 16. Examples of Safety Containers.

Disinfection of Aseptic Safety Cabinets

Various disinfectants are available; ethylene oxide and β-propiolactone are often preferred because they are gaseous at room temperature. However ethylene oxide is explosive unless mixed with e.g. CO_2 Sykes (1965); and toxic, while β propiolactone is, in addition, reputed to be carcinogenic. We therefore prefer to vaporize aqueous solutions of either formaldehyde (Darlow, 1958), or glutaraldehyde (Evans and Harris-Smith, 1970). Both are very effective provided that the relative humidity is maintained above 75%. Formaldehyde, being cheaper, is normally used for disinfecting laboratories and other large spaces, but glutaraldehyde is preferred for safety cabinets and delicate apparatus because it is much less liable than formaldehyde to polymerize on surfaces. For each ft^3 of space we vaporize 30 ml of a 10% solution of glutaraldehyde which is left to disinfect overnight.

Operating Aseptic Safety Cabinets

The main difference in operating this, compared with other types of safety cabinet, lies in the steps taken to clear the cabinet of infection before and after culture manipulations. This clearance is done by ventilating, with the UV lamps switched on, for about 30 min after all necessary cultures and materials have been placed in the cabinet. When work in the cabinet is finished, this period of ventilation and irradiation is repeated before the cultures are brought out. Otherwise, the usual rules apply; regular testing of the cabinet's safety (i.e. air flow rate, pressure differential, filter efficiency), and UV tubes should be regularly checked as described by Morris (1972). When working with established pathogens, personal gloves are worn as an added protection, and disinfected after use. Finger nails should always be kept short, and sharp objects such as wrist watches and engagement rings should be removed.

Design of Enclosures for Apparatus which may Release Aerosols

The general principles applicable to the design of aseptic safety cabinets for enclosing established pathogen culture apparatus, have been previously described (Harris-Smith and Evans, 1968). These principles can be extended to other pieces of laboratory equipment; Fig. 12 shows a prototype enclosure for a Hi-speed M.S.E. centrifuge,* with access through a disinfectant liquid lock. Figure 13 shows a similar enclosure for an M.S.E.

* Measuring & Scientific Equipment Ltd., 14–28 Spenser Street. London, S.W.1, England.

"Medium" centrifuge (H. M. Darlow, pers. comm.) with a flap valve controlling the ventilation air inlet. Figure 14 shows a continuous flow centrifuge* enclosure which was designed to work in conjunction with the POMEC shown in Fig. 15.

The principles exemplified by these special designs may be summarized as follows:

1. The interior of the cabinet should be smooth, free from crevices and pinholes, and without sharp edges which could damage gloves.
2. There should be an accident well of capacity sufficient to retain all the liquid contained in the cabinet, which could be accidentally spilled.
3. The comfort of the operator is important; if operated by a seated worker, the front should have a 10° slope, but for a standing worker, a vertical front is adequate. (For a full discussion of the ergonomic aspects, see Doxie and Ullom, 1967.)
4. The position of the gloves is important; all necessary parts of the cabinet should be within reach. The oval glove ports shown in Figs 11, 12, 14 and 15 were designed by us to give increased freedom of movement, but unfortunately, at present, there is insufficient demand for them to interest manufacturers. The gloves must not be liable to damage by the apparatus, e.g. pumps, burners, centrifuges, etc.
5. A disinfectant lock should be provided if material is to be removed before the cabinet can be disinfected.
6. The flow of ventilation air through the cabinet should be sufficient to give one air change per minute, and so directed that all parts of the cabinet are cleared. The resistance of the filters should be such that the pressure within the cabinet is between $\frac{1}{2}$ and 1 in. w.g. less than that outside it.

Any cabinet should be designed for an operator, not against him; there is no point in equipping a laboratory with cabinets that make life unnecessarily difficult for the operator. A tired or frustrated operator is himself a hazard and is liable to take "short cuts" that might prove disastrous to the cultures, to himself, and, worse still, to others.

Although the other types of cabinet discussed are quicker to use than aseptic safety cabinets, this apparent disadvantage is more than compensated for by the safety offered to the operator, other laboratory personnel, and the avoidance of culture contamination.

There is an eighteenth century calculating machine in the Science Museum which, it was claimed, could be worked "without charging the memory, disturbing the mind or exposing the operator to any uncertainty". In matters of safety, this is the essence of good design.

* Sharples Centrifuge, Pensalt Ltd., Tower Works, Doman Road, Camberley, Surrey, England.

References

ANON (1970). Airborne transmission of smallpox. *WHO Chron.* **24,** 311.

BARBIETO, M. S. & TAYLOR, L. A. (1968). Containment of microbial aerosols in a microbiological safety cabinet. *Appl. Microbiol.,* **16,** 1225.

BOURDILLON, R. B., LIDWELL, O. M. & LOVELOCK, J. E. (1948). Studies in Air Hygiene. *Med. Res. Coun. Spec. Rept.* No. 262 London: H.M.S.O.

CHATIGNY, M. A. & CLINGER, D. I. (1969). In *An Introduction to Experimental Aerobiology* (R. L. Dimmick and A. B. Ackers, eds) p. 218. New York: Wiley Interscience.

CORIELL, L. L. & MCGARRITY, G. J. (1968). Biohazard hood to prevent infection during microbiological procedures. *Appl. Microbiol.,* **16,** 1895.

CORIELL, L. L. & MCGARRITY, G. J. (1970). Evaluation of the Edgegard laminar flow hood. *Appl. Microbiol.,* **20,** 474.

CRUICKSHANK, R. (1969). *Medical Microbiology* (R. Cruickshank, J. P. Duguid and R. H. A. Swain, eds) 11th edition p. 797. Edinburgh: E. & S. Livingstone Ltd.

DARLOW, H. M. (1958). The practical aspects of formaldehyde fumigation. *Month. Bull. Minist. Hlth,* **17,** 270.

DARLOW, H. M. (1967). The design of microbiological safety cabinets. *Chemy. Ind.,* 1914.

DARLOW, H. M. (1969). Safety in the microbiological laboratory. In *Methods in Microbiology* (J. R. Norris and D. W. Ribbons, eds) Vol. 1, pp. 169–204. London and New York: Academic Press.

DARLOW, H. M., HARRIS-SMITH, R. & EVANS, C. G. T. (In preparation.) An apparatus for growing pathogenic microorganisms. The POMEC: II Physical and biological testing.

DOXIE, F. T. & ULLOM, K. J. (1967). Human factors in designing controlled ambient systems. *West. elect. Eng.,* **11,** 24.

ELSWORTH, R., MORRIS, E. J. & EAST, D. N. (1961). The heat sterilization of spore infected air. *Chem. Eng., Lond.,* **157,** A 47.

EVANS, C. G. T. & HARRIS-SMITH, R. (1970). The POMEC, an apparatus for growing dense cultures of pathogenic microorganisms. In *Automation, Mechanization, and Data Handling in Microbiology*. The Soc. Appl. Bact. Tech. Series, **4,** (A. Baillie and R. J. Gilbert, eds) pp. 137–149. London and New York: Academic Press.

HARRIS-SMITH, R. & EVANS, C. G. T. (1968). The Porton Mobile Enclosed Chemostat (POMEC). In *Continuous Cultivation of Microorganisms*, Proceedings of the 4th Symposium Prague, June 17–21, 1968, pp. 391–410. Prague 1969: Academia.

HARRIS-SMITH, R., PIRT, S. J. & FIRMAN, J. E. (1963). A ventilated germ-free cabinet for the microbiological laboratory. *Biotechnol. Bioengng.*, **5,** 53.

HENNESSEN, W., BONIN, O. & MAULER, R. (1968). Zur Epidemiologie der Erkrankung von Menschen durch Affen. *Dt. med. Wschr.*, **93,** 582.

MACKIE, T. J. & MCCARTNEY, J. E. (1948). *Handbook of Practical Bacteriology*. 8th Ed. p. 200. Edinburgh: E. & S. Livingstone.

MCDADE, J. J., SABEL, F. L., AKERS, R. L. & WALKER, R. J. (1968). Microbiological studies on the performance of a laminar airflow biological cabinet. *Appl. Microbiol.*, **16,** 1086.

MORRIS, E. J. (1972). The practical use of ultraviolet radiation for disinfection purposes. *Med. Lab. Technol.*, **29**, 41.

STAAT, R. H. & BEAKLEY, J. W. (1968). Evaluation of laminar-flow microbiological safety cabinets. *Appl. Microbiol.*, **16**, 1478.

SYKES, G. (1965). *Disinfection and Sterilisation.* 2nd Ed. London: Spon Ltd.

TREXLER, P. C. (1957). Improvements relating to biological apparatus. Brit. Pat. No. 770,491.

WHITE, P. A. F. & SMITH, S. E. (1964). (eds) *High Efficiency Air Filtration.* London: Butterworths.

WILLIAMS, R. E. O. & LIDWELL, O. M. (1957). A protective cabinet for handling infective material in the laboratory. *J. clin. Path.*, **10**, 400.

Aerosol Sampling

F. A. Dark and G. J. Harper

Microbiological Research Establishment, Porton Down, Salisbury, Wiltshire, England

There are many reasons for wanting to sample the atmosphere for the presence of particles, especially those of biological origin. This may enable one to establish the cause and spread of infectious diseases, and knowledge of the presence of certain biological hazards would be invaluable in the pharmaceutical, food producing and brewing industries. When planning air sampling programmes for investigating these problems, there are 3 measurements that are of the utmost importance: (1) concentration of organisms in the air, (2) the size of the particle in which the organisms are contained, and (3) the change in concentration of airborne organisms with time.

1. *Concentration.* With certain diseases, only small numbers of organisms are required to initiate infection *via* the airborne route, so it is essential to be able to sample a large volume of air and concentrate the microorganisms into a small volume of liquid for subsequent identification. When the object of air sampling is to monitor the air in clean handling areas, then clearly the ability to detect the presence of low concentrations of microorganisms is of primary importance.

2. *Particle size.* The size of the particles carrying the microorganisms is of importance in predicting its penetration and retention in the respiratory tract as well as the maintenance of viability during airborne travel. Detailed knowledge of the size of the airborne particles is also essential for the design of air filtration systems.

3. *Detection of changes with time.* The ability to detect changes with time in the concentration or type of microorganisms in the aerosol can be used to pinpoint cause and effect relationships.

The aims of an air sampling programme will dictate the type of air sampling device to be used. At first sight the existence of a large number of sampling devices makes the choice difficult, but few principles are involved in recovering microorganisms from the air and, when these are understood, the choice of sampling equipment becomes easier.

In this chapter an attempt is made to cover some of the basic aspects of air sampling and describe in more detail sampling devices which have been developed over many years at the Microbiological Research Establishment (MRE), Porton. We will then discuss the advantages and disadvantages of some of the samplers in assessing a specific aerosol problem.

The Basic Principles of Air Sampling

Air sampling devices can be split into three groups: 1. Inertial collectors, 2. Filters, and 3. Electrostatic and thermal precipitators.

Inertial collectors

When designing this type of sampler, use is made of the knowledge that particles moving in an air stream have an individual inertia and may be deflected from their original path on to a surface (to be trapped by **impaction**), or into a liquid (to be trapped by **impingement**). The inertia of the particles depends upon their size and weight and knowledge of this fact is used to design particle size samplers. Some examples of inertial samplers are given below.

The settle plate. This is probably the most simple sampling device. Usually it consists of an open Petri dish containing Nutrient Agar on to which particles which have an inertia owing to gravity are allowed to settle. Subsequent incubation produces colonies corresponding to these biological particles. This method is best suited for still air conditions as it tends to select very large particles only; the greater the air movement the larger are the sizes selected. Sometimes an open container of liquid is used, the liquid being subsequently assayed for its biological content.

Impingers. These have a jet or slit which, when air is passed through, increases the velocity of the air and hence the inertial velocity of the particles moving in it. The sampler contains liquid through which the air is forced, the particles being extracted and suspended in the violently agitated liquid which is eventually assayed for microorganisms.

The impactor. This uses the principle of the impinger but the particles are directed at a solid surface.

Air centrifuges. These remove particles from the air by centrifugal force, normally on to the inner surface of a rotating cone. The cyclone sampler uses this principle. The geometry of the sampler and the positioning of the inlet tube imparts a rotary action to the air and centrifuges out the particles on to the walls of the apparatus from which they are washed by a liquid.

Filters

Filtration is probably the most commonly used means for particle sampling. By varying the type of filter, its size, and the length of time of sampling, very wide application of this technique is possible. The theory of filtration depends on several factors:

1. "Membrane" type filters retain particles by direct action when these are larger than the effective pore size of the filter material.
2. Inertial filtration takes place in the fibrous type filter, in which the inter-spaces are larger than the particles and efficiency of filtration depends on contact between particles and filter fibres within the material.
3. Electrostatic forces contribute largely to the efficiency of some types of filters.
4. Other methods rely on diffusion and settling within the material.

Most of the filters designed for air sampling are very well described by the various manufacturers and are not considered in detail here. The effect of excessive drying on the viability of most vegetative microorganisms restricts the use of filters as air sampling devices to the collection of resistant cells such as bacterial spores. The high pressure drop across the filter is another factor which limits the application of filtration methods.

Electrostatic and thermal precipitators

Although electrostatic forces play some part in many of the mechanical sampling devices, in the electrostatic collector this force is the only one used. Particles are given an electrical charge by one of many ways (friction, ionization in flame, ionizing radiation, high voltage corona discharge) and the charged particles are collected by being rapidly attracted to an electrode of opposite polarity.

Thermal precipitators depend for their efficiency on the fact that particles acquire a thermophoretic force in a temperature gradient and can be made to deposit on to a slide for microscopic examination. This method is very efficient for the collection of small (sub micron) particles and for use with the electron microscope, but unsuitable for the recovery of viable biological particles. As the precipitation velocity is low, they are useful only at very low flow rates.

Sampling Apparatus

The following 6 sampling devices have been developed and used successfully at MRE over many years, and when used either singly or in selected groups they should give information adequate for solving most problems connected with aerosol sampling.

Porton raised impinger (May and Harper, 1957)

This compact and inexpensive apparatus (Fig. 1) is a modification of a liquid impinger first described by Greenburg and Smith (1922) and later developed by Rosebury (1947) and Henderson (1952). The sampling fluid can be plated on to different media and by suitable dilution techniques it can accomodate extreme ranges of airborne concentration. Its particle retention efficiency is very high for particles down to 0·5 μm diam. Thus it is efficient for collecting virus aerosols which are not usually dispersed in sub-micron particles. It measures the total number of organisms in a sample and not merely the number of particles. It is, moreover, its own constant flow meter. and is easily cleaned and sterilized by autoclaving. Its main disadvantage is that, owing to its high impingement velocity, some sensitive organisms may be killed during collection.

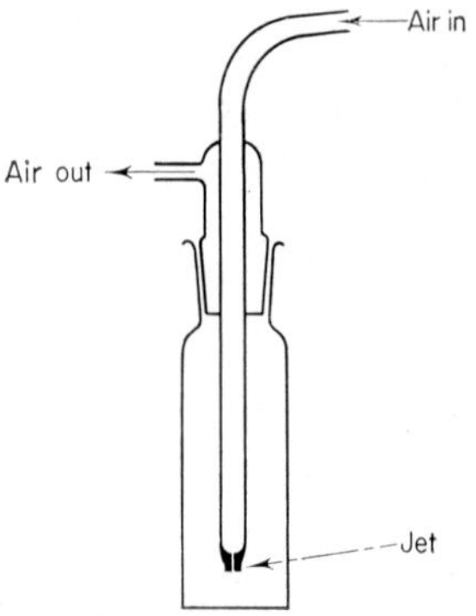

FIG. 1. Porton raised impinger.

The sampler depicted in Fig. 1 is of all glass construction and the jet is a short length of capillary tubing which acts as its own critical orifice when the pressure drop across it exceeds half an atmosphere and the flow then attains sonic velocity (Druett, 1955). Any further increase in suction cannot increase jet velocity. It is usually designed to work at 11 or 25 l/min. This sampler was recommended as a standard reference sampler by an expert panel (Brachman *et al.*, 1964).

Cascade impactor (May, 1945)

The cascade impactor (Fig. 2) is used to collect airborne particles on to microscope slides. The sample is split into a series of four progressively smaller particle size ranges by decreasing the size of the jet at each stage.

As the mass median size of the particles collected at each stage is known or can be determined, then measurements of the relative amounts collected on each slide gives the size distribution of the sample tested. The particles impinged on to the slides may, after suitable staining if necessary, be

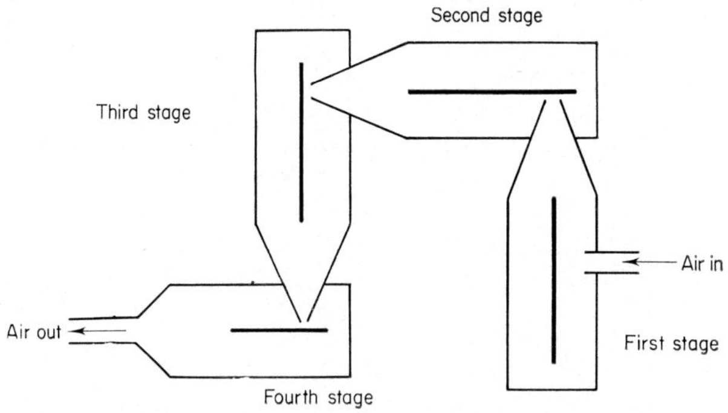

FIG. 2. Cascade impactor. Broad lines, microscope slides the size of which are given in the text.

examined directly under the microscope, or washed off and plated out on to nutrient medium for counting and identification. This latter method is only applicable if the organism is resistant to drying.

Figure 2 shows the general layout of the cascade impactor. It consists of 4 stages. Microscope slides (3 × 1 in. for the first stage and $1\frac{1}{2}$ × 1 in. for the subsequent stages) are shown as thick lines, and they are held in place against the runners by springs (not shown in the diagram). The jets are of progressively smaller cross section. The first is 19 × 7 mm and the lower edge projects to within 1 mm of the surface of the slide, the upper edge ends 6 mm from the slide. Jet No. 2 is 14 × 2 mm and is chamfered at 45° with the lower edge nearly touching the slide. Jets Nos. 3 and 4 are 14 × 1 mm and 14 × 0·5 mm respectively. The shaping of jets 1 and 2 forces most of the air along the slide towards the next stage. The sampling rate is 17·5 l/min. The size graded fractions are: Stage 1, 6–20 μm; Stage 2, 2–6 μm; Stage 3, 1–3 μm, and Stage 4, 0·5–1·5 μm.

Multistage liquid impinger (May, 1966)

The multistage sampler is constructed of glass and collects airborne particles into liquid at a fairly high sampling rate. It may be obtained in three different sizes but in general we have found that the model sampling at 50 l/min to be the most useful. By its particle size distribution, the impinger is intended to simulate some of the final details of the human respiratory tract—see p. 167. This sampler is robust and easily transportable. It concentrates fairly dilute aerosols by utilizing prolonged sampling times without great loss of viability, as the impingement is gentle and evaporation low. The sampler as shown in Fig. 3 is constructed of Pyrex glass. The three chambers are numbered 1 to 3 from top to bottom. The air inlet to chamber 1 has a smooth bell-mouthed entry and a bore of 15 mm. The connecting tube from chamber 1 to chamber 2 is a straight tube of 10 mm bore and also

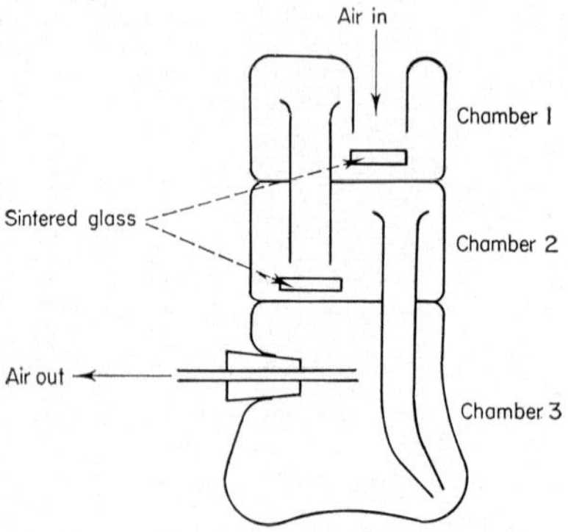

FIG. 3. Multistage liquid impinger.

has a bell mouth. The tube from chamber 2 to chamber 3 is also of 10 mm bore and bell-mouthed but is curved and tapers gradually to a nozzle with an internal diameter of 3·3 mm. Circular discs of coarse sintered glass are fixed below the entry tube and the connecting tube from chamber 1 to chamber 2. These are twice the diameter of the corresponding tubes, are fixed 3/8th of the diameter below the ends of the tubes and are constantly wetted by the liquid in the chamber. Access holes to each chamber are sealed with rubber bungs, the lower bung being fitted with a tube for connecting to a suitable vacuum pump. A constant flow critical orifice is fitted to this tube.

The cyclone separator (Errington and Powell, 1969)
The cyclone separator is a modification of the large industrial cyclones which are used to remove dust. To make it suitable as a biological sampling device it has been fitted with a liquid spray. The liquid spray is thrown on to the walls and continuously washes down the deposited particles through the tail pipe into a separate replaceable container. It has been made in two different sizes, a large model constructed of Perspex operating at 360 l/min and a smaller model made of stainless steel operating at 75 l/min. Suitable critical orifices are used to regulate the flow of the air through the samplers. The cyclone is suitable for sampling dilute aerosols as it achieves a high degree of concentration, it is robust, reliable and will provide a long

sequence of samples. Figure 4 shows a cross section of the device. In the small 75 l/min model the main cylindrical part is $\frac{1}{2}$ in. in diameter, the throat is $\frac{1}{4}$ in. sq. section and leads to the volute which is milled into the centre square section. The inlet is an annular gap between two plates and there is a fine tube at the centre through which the scrubbing liquid is fed.

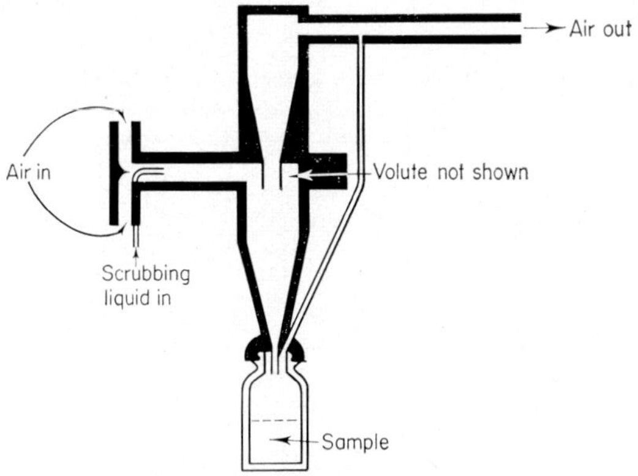

FIG. 4. Cyclone separator.

Samples are collected into *bijou* bottles screwed into the lower end of the sampler. The larger cyclone has dimensions twice that of the smaller version and the proportions are the same except for the throat which is relatively smaller. Using the large sampler it is possible to concentrate the particles contained in 500 l of air into 1 ml of liquid.

Electrostatic precipitator (Morris, Darlow, Peel and Wright, 1961)

There are three variants of this sampler, all having an outside Perspex tube. In the standard form, the tube is lined by a glass cylinder which is coated with Nutrient Agar and connected to the earth terminal. Air is led continuously through the cylinder whilst a positive potential is applied to a central electrode. Charged particles are deposited on to the agar surface. After a suitable sampling time, the whole inner cylinder is removed and incubated. Counts of the colonies produced on the agar give the total number of viable **particles** in the sample of air drawn through the apparatus. The first variation of this apparatus was a syringe attachment to draw in small measured volumes of air instead of the continuous supply of air used in the standard version. With the third model, the internal cylinder contains liquid instead of agar and the cylinder is rotated on a horizontal axis. The liquid can be assayed in any desired medium and, unlike the model impinging directly on to agar, this variant yields estimates of the total number of viable **organisms.**

A collecting efficiency approaching 100% is readily obtained using this type of sampler and it is probably the only practical method of sampling for submicron particles. The standard version of the apparatus is shown in Fig. 5. It has screw cap ends to the outside cylinder of Perspex, one cap carrying the air inlet and the other a pointed stainless steel electrode, an air outlet and an adjustable earth terminal which may be moved to contact the agar surface of the internal glass cylinder. The internal glass cylinder (2 × 12 in.) is coated on the inside with a layer of Nutrient Agar. The tension

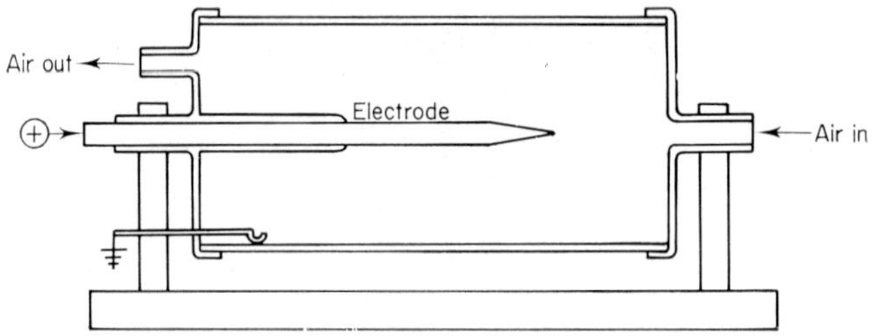

FIG. 5. Electrostatic precipitator.

applied to the central electrode is varied with the air flow so that maximum collection efficiency is obtained. For example at 10 l/min flow, 8 kV tension is used; at 5 l/min, 14 kV is used. In the small volume variant of the sampler, the attached syringe has a total volume of 150 ml, the electrode is mounted in the piston of the syringe and moves along the length of the tube as the sample is drawn in. With the third version, the internal glass collecting cylinder has 1 in. intrusions at each end to prevent the sampling liquid spilling. The cylinder is held in place in the outside Perspex tube by means of a rubber moulding and the whole is rotated by an electric motor. The collecting fluid (5–10 ml) forms a thin layer on the inside surface and continuously washes down the walls.

Slit sampler with interval timing

The slit sampler as used at MRE is a modification of the Casella slit sampler (C. F. Casella & Co. Ltd) which was devised by Bourdillon, Lidwell and Thomas (1941). In the original and widely used instrument, air is sucked in through a slit on to the surface of Nutrient Agar in a Petri dish which is rotated continuously at some predetermined speed so that the organisms are distributed over the surface of the medium. With the modified version used at MRE, the plate is rotated in a series of angular steps each of 6°. The length of time the sampler impinges during each step is regulated by an electronic timer which can be set to move on at intervals of 15, 60, 120 sec, etc. When operating in this way, up to 59 traces can be accommodated in a single Petri dish. The time intervals could of course be changed to suit individual requirements. A further modification (at present in the development stage only) seeks to provide an additional timing circuit so that both the sampling periods of each segment and the time interval between samples will be controlled. For instance one might want to take 15 sec samples of air once every 30 min or say 30 sec samples ever 15 min.

The Choice of Sampling Systems

When selecting a sampler or samplers for use in any particular situation a number of factors have to be considered. Is the sampling programme designed to search for a particular organism or group of organisms? Is the object to measure the concentration of all organisms present in the atmosphere? Are the likely concentrations in the air going to be high or low? Are the numbers of **cells** or the numbers of **particles** of primary importance? Clearly all such questions and perhaps more have to be asked before starting to sample the air.

If the object is to monitor air for the presence of low concentrations of organisms, then a large volume sampler such as the cyclone or the Litton large volume air sampler (Litton Systems Inc, Minneapolis, Minnesota, U.S.A.) will be needed. If the suspected concentrations are very low, then the sampling liquid may have to be recirculated through the sampler to obtain sufficient cells for assay or detection. For most kinds of cells the use of May's multistage sampler is a good choice. The sampling liquid can be used to inoculate a range of culture media and can be diluted to cope with high cell concentrations. The additional information on the particle size characteristics of the airborne cells is of considerable value.

Time discrimination is best dealt with by the use of some form of slit sampler, and again the choice will depend on the expected concentration range to be sampled. The cyclone and Litton samplers have also been useful when used in this way particularly when sampling in low concentrations.

Particle size information can be obtained by using the cascade impactor, the multistage sampler or the Andersen (1958) sampler. All these samplers collect at relatively low flow rates and may need to be operated for very long periods to accumulate enough cells for a satisfactory assessment. The Andersen sampler can be used both as a cell and a particle collector and this enables a measurement of the number of cells in particles of given size ranges to be estimated. The cascade impactor can be used for microscopical sizing of the collected particles, or if some robust cell, resistant to desiccation, such as a bacterial spore is present in the cloud, the mass on each stage can be assessed and the information used to estimate the Mass Median Diameter (MMD).

Although any sampling device that can efficiently remove small particles from the air can yield material for examination, it must be remembered that the object of sampling air for its microbial content is to obtain quantitative recovery of the cells in as far as possible the same state of viability as they exist in the atmosphere. It has been shown that sampling devices which incorporate collection into a liquid or on to a moist surface give the highest recovery of viable bacteria. The recovery of most bacteria or viruses from systems involving collection on dry filters, in itself an **efficient collecting system**, results in **low recovery** of **viable cells.**

Cells that have been held in the airborne state and subjected to the trauma of sampling, often exhibit increased sensitivity to many inimical factors when compared with cells that have not been so stressed. Care, therefore, needs to be taken in the choice of collectors, assay methods and materials. The use of inhibitors such as dyes or antibiotics in culture media should be avoided whenever possible. If selective media have to be used, in air sampling programmes, then the inhibiting or selective factors should be checked with stressed, preferably airborne, cells. The choice of the most suitable collecting fluid is often governed by the availability of assay facilities. The ideal collecting fluid should preserve the *status quo* of the organism at the time of collection. However every effort should be made to assess air samples as soon as possible after collection.

Another factor of considerable importance in planning an air sampling programme is the provision of adequate vacuum to operate the sampling equipment. Most samplers need a control device to meter the volume of air being sampled. This can be achieved by the use of flow meters, pressure gauges or manometers but by far the simplest and most reliable way of controlling the flow rate is by the use of critical orifices. The introduction by Druett (1955) of low pressure drop orifices has reduced considerably the vacuum needs for most sampling systems. A wide range of vacuum pumps operated by electric motors is available commercially and in choosing a pump and motor, the power requirements of the system must be borne in mind. There is little point in having an efficient pumping system and an inadequate power supply. These considerations do not apply to certain samplers such as the Litton large volume sampler or some of the electrostatic samplers—these samplers have their own built in vacuum and HT systems.

Care must be exercised when sampling air suspected of containing microbes pathogenic to man, animals or plants. The effluent air from samplers and pumps will possibly contain small numbers of viable cells that have passed through the sampler and the air should be passed through an efficient filter, incinerator or other satisfactory sterilizing system before being discharged to the atmosphere.

References

ANDERSEN, A. A. (1958). New sampler for the collection, sizing and enumeration of viable airborne particles. *J. Bact.*, **76,** 471.

BOURDILLON, R. B., LIDWELL, O. M. & THOMAS, J. C. (1941). A slit sampler for collecting and counting airborne bacteria. *J. Hyg. Camb.*, **41,** 197.

BRACHMAN, P. S., EHRLICH, R., FUCHENWALD, H. F., CABELLI, V. J., KETHLEY, T. W., MADIN, S. H., MALTMAN, J. R., MIDDLEBROOK, G., MORTON, J. D., SILVER, I. H. & WOLFE, E. K. (1964). Standard sampler for assay of airborne microorganisms. *Science, N.Y.*, **144,** 1295.

DRUETT, H. A. (1955). The construction of critical orifices working with small pressure differences and their use in controlling air flow. *Br. J. ind. Med.*, **12,** 65.

ERRINGTON, F. P. & POWELL, E. O. (1969). A cyclone separator for aerosol sampling in the field. *J. Hyg. Camb.*, **67,** 387.

GREENBURG, L. & SMITH, G. W. (1922). *A New Instrument for Sampling Aerial dust.* US Bureau of Mines Reports of Investigations. Serial No. 2392.

HENDERSON, D. W. (1952). An apparatus for the study of airborne infections. *J. Hyg. Camb.*, **50,** 53.

MAY, K. R. (1945). The cascade impactor: an instrument for sampling coarse aerosols. *J. scient. Instrum.*, **22,** 187.

MAY, K. R. (1966). Multistage liquid impinger. *Bact. Rev.*, **30,** 559.

MAY, K. R. & HARPER, G. J. (1957). The efficiency of various liquid impinger samplers in bacterial aerosols. *Br. J. ind. Med.*, **14,** 287.

MORRIS, E. J., DARLOW, H. M., PEEL, J. F. H. & WRIGHT, W. C. (1961). The quantitative assay of mono-dispersed aerosols of bacteria and bacteriophage by electrostatic precipitation. *J. Hyg. Camb.*, **59,** 487.

ROSEBURY, T. (1947). *Experimental Airborne Infection.* Baltimore: Williams & Wilkins Co.

Laminar Flow as Applied to Bacteriology

M. J. Yalden

W. H. S. (Pathfinder) Limited, Solent Road, Havant, Hampshire, England

The use of laminar flow clean air equipment in microbiology is a relatively new introduction, but it has already found wide application. The main purposes for which such equipment is required are described below.

Where work has to be carried out under a controlled environment free from air-borne bacteria, particularly pathogens, the laminar flow unit, as opposed to the glove box (see p. 21), provides an advantage in that the working space is continually purged with clean air which moves in a non-turbulent flow pattern. This minimizes the risk of cross contamination between one operation and another carried out in the same box. In a glove box where the flow of clean air is at a relatively low rate and where it enters one or more corners with a turbulent flow pattern, then any particulate contamination generated within the space is liable to be mixed uniformly with the air and transferred to other portions of the controlled environment.

The laminar flow cabinet, on the other hand, may be available in its horizontal flow form with a completely open front, making operator use more easy and more comfortable and with improved accessibility. It is also more efficient. The principle of operation is shown in Fig. 1, and two types of cabinet are illustrated in Fig. 2 and Fig. 3.

Filters normally used in such a laminar flow cabinet are efficient to 99·997% on BS 3928 (Sodium flame test) which is an aerosol with a median of 0·4 μ. Some cabinets are made with lower efficiency filters, but in view of the small difference in cost between the most efficient filters and other High Efficiency Particulate Filters (Hepa filters), it is worth having the additional safety. Construction of the cabinets is another factor influencing the efficiency of their operation. This should be of a non-noise generating type, constructed to restrict the noise level and vibration. The most general type of construction embodying these principles is one where a melamine finish on a chipboard core is used, care being taken to ensure that no possibility exists for the growth of bacteria within the core material. Cabinets are manufactured of easily cleanable and chemically resistant materials

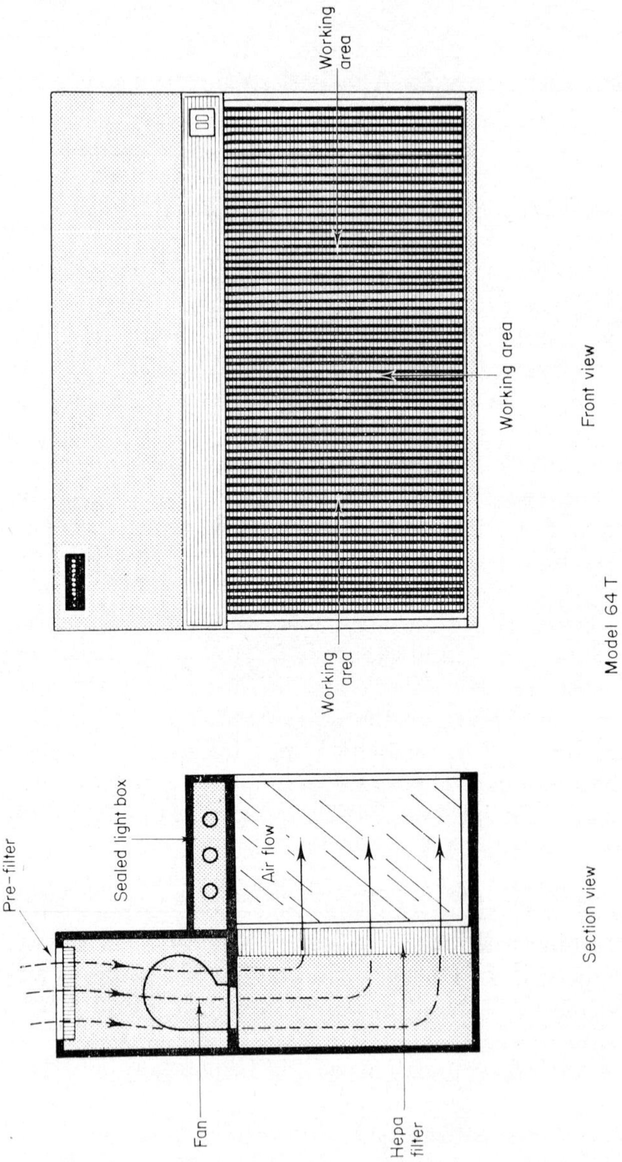

FIG. 1. Principle of operation of horizontal laminar flow air cabinet.

FIG. 2. The Model 64 T horizontal flow work station for mounting on existing work bench.

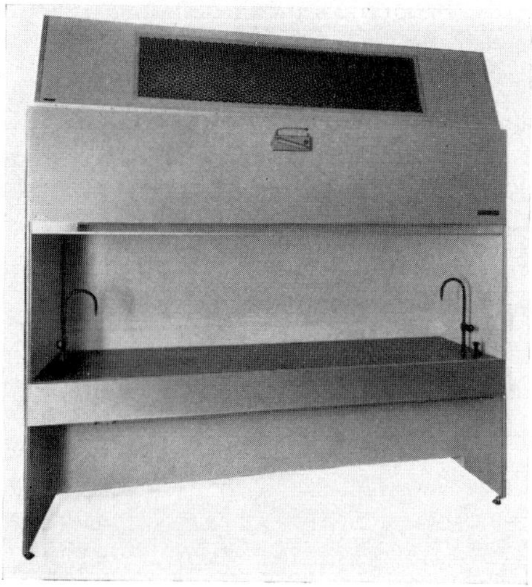

FIG. 3. Vertical flow work station for bacteriological testing.

such as melamine laminate or stainless steel, operating from a normal 13A socket outlet.

An example of the effectiveness of an open-front horizontal laminar flow cabinet has been demonstrated whereby a gnotobiotic animal was kept for a period of eight days in a cabinet under germ-free conditions (Dr. D. Blackmore, pers. comm.).

For carrying out operations in a controlled environment where toxic bacteria or pathogens are being handled and where there is a distinct danger to the operator and to the outside laboratory, then a different type of cabinet is required. This requires a filter on the exhaust air duct as well as on the air in-put. Such cabinets are also available in a laminar flow form and this is usually a vertical flow. The air passes through prefilters and then downwards through a high efficiency air filter in a vertical laminar flow direction, out through a perforated work top (which may only be partly perforated if required), and exhausted through another Hepa filter to the outside atmosphere. Screens may be incorporated in the front of the cabinet and glove ports used where there is a risk of transfer of contaminants by contact. If, however, the pathogens are mainly air-borne, then the laminar flow system is extremely effective, providing the in-put air is balanced with the exhaust volume, to ensure an air flow pattern in the cabinet which does not result in any of the contaminated air being expelled outwards into the room.

A particular type of cabinet, a "pathological cabinet" for bacteriological use, has been developed for handling tubercular specimens and carrying out other potentially dangerous manipulations. In these circumstances contamination from the outside laboratory and its effect on the work is of far less importance than the possible danger to the operator. Therefore, in these circumstances, there is no filtered air input but the cabinet is constructed in such a way that the exhaust air is filtered. A Hepa filter is ideal for this purpose but there is risk when the filter requires changing that contamination from it could infect the operator. Therefore a number of possible alternatives are to be considered. The Hepa filter is subjected to continual UV radiation, although this is considered in some quarters to have a deleterious effect. The exhausted contaminated air is passed through a heated section in order that any active constituents may be destroyed.

The Management of Laboratory Discard Jars

Isobel M. Maurer

Disinfection Reference Laboratory, Central Public Health Laboratory, Colindale Avenue, London NW9 5HT, England

Equipment which has been contaminated in a bacteriological laboratory must be rendered safe to handle before it is either thrown away in dustbins or washed and used again. Most of it is discarded into bins for autoclaving but pipettes treated in this way are often difficult to clean. Although, in a few laboratories, pipettes are regularly autoclaved and are cleaned satisfactorily, it is more usual to place them along with other items such as tubing in a jar containing a disinfectant solution, for decontamination before handling by the staff who are responsible for the washing.

Choice of a Disinfectant

No one proprietary disinfectant is ideal for use in discard jars. Some very popular products, such as quaternary ammonium compounds, are a poor choice for laboratory use because of their narrow antibacterial spectrum. Those which are generally selected as having a wide spectrum are either phenolics or solutions of sodium hypochlorite. Of the first, the black and white fluid phenolics are effective decontaminants but are difficult to remove from the surface of glassware. Clear soluble phenolic fluids of the lysol type are a good choice; suitable brands are Clearsol, Hycolin, Stericol and Sudol. Lysol, as such, has been used in many laboratories but is becoming increasingly difficult to buy. It is not recommended because of its caustic nature. The principal advantage of phenolic disinfectants, apart from their wide spectrum, is their low level of inactivation by organic material, although this is not true of chloroxylenol. A disadvantage of phenolics is absorption by rubber on prolonged contact. Phenolics are recommended by the Department of Health and Social Security for use in laboratories where there may be contamination by tubercle bacilli.

Solutions of sodium hypochlorite are normally much cheaper than phenolic disinfectants. Hypochlorites are recommended for discard jars where contamination by viruses is to be expected. The limited amount of information available suggests that hypochlorites are effective against all types of viruses, while this is not true of phenolics. The principal disadvantage of hypochlorites is the readiness with which they are inactivated by organic material. Other disadvantages are ineffectiveness against tubercle bacilli and the fact that metals, including some stainless steels, are corroded by all other than those expensive preparations which are specifically described as non-corrosive. The lack of wetting power shown by a hypochlorite in contact with a highly polished surface may be overcome by the addition of a compatible detergent, of which details are given below. Failure to wet a surface, may mean failure to disinfect it.

Disinfectants and detergents

Microorganisms may be protected from the action of a disinfectant by the presence of organic material. The addition of a detergent, at a concentration of *c.* 1·0% (v/v) will aid penetration of a heavy load of contaminated material in a discard jar. In mixing a detergent with a disinfectant, it is essential that compatible products be chosen. Phenolics and hypochlorites are both anionic and are compatible with anionic or non-ionic detergents. An anionic disinfectant will be inactivated by the presence of a cationic detergent and microorganisms are likely to survive in such a mixture. In the case of Gram negative organisms there may be not only survival but also growth.

Concentration of disinfectants

All disinfectants are inactivated in varying degrees by numerous substances. Organic material such as blood, serum, pus, urine, faeces, media, scraps of food or a heavy load of microorganisms may inactivate a disinfectant, as well as protecting some of the organisms from contact with it. Other inactivating substances include incompatible detergents, soap, hard water, cork, plastics and cellulose products. Inactivation may be complete or partial depending on the nature and concentration of the disinfectant and on the kind and quantity of inactivating material. Thus, a recommended concentration of a disinfectant for use in a discard jar can never be guaranteed to be effective. It is essential for laboratory staff to arrange for regular in-use testing to be carried out as a matter of routine and immediately the nature of the work in the laboratory changes. The simple in-use test of Kelsey and Maurer (1966) involves removing at the end of the day a 1 ml sample of the disinfectant from the discard jar and adding it to 9 ml diluent. In the case of phenolic disinfectants, quarter-strength Ringer's solution is a suitable diluent, but the bacteriostatic non phenolic disinfectants should be diluted with Nutrient Broth containing a neutralizer. Rubbo and Gardner

(1965) give a useful list of neutralizers for many kinds of disinfectants and a manufacturer will usually name a neutralizer for his own product. The action of the neutralizer used should be checked by the method described in British Standard 3286 (Anon, 1960). A sample (Fig. 1) of the disinfectant diluted in Ringer's solution or Neutralizer Broth is withdrawn and 10 drops are placed on the dried surface of Nutrient Agar contained in 2 Petri dishes.

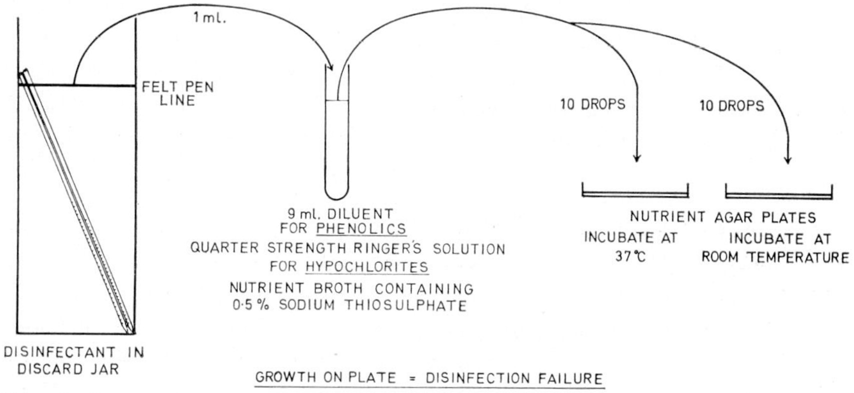

FIG. 1. In-use test.

The plates are incubated up to 72 h, one at room temperature and the other at 37° or the optimum temperature for the organisms which may be expected in the discard jar. Growth on either or both plates indicates a failure in disinfection.

Unfortunately the importance of using a neutralizer in a test of this kind has not always been appreciated. On occasions, neutralizers have not been used and the absence of growth on the subculture plates has been wrongly interpreted as an absence of viable organisms. Some non-phenolic disinfectants are highly bacteriostatic and will, in spite of dilution, prevent the growth of surviving microorganisms. A neutralizer must be used to counteract this bacteriostatic action and make it possible for surviving organisms to grow and form colonies on the sub-culture plates. It is also necessary to be sure that the neutralizer does not inhibit the growth of the organisms. The checking of the neutralizer by the British Standard (Anon, 1960) method is thus essential to ensure the choice of a type and a concentration of a neutralizer suited to the particular disinfectant at the concentration used in the discard jar. Some published work describing tests with particular disinfectants has been misleading, as the authors have not understood the need for the use of a suitable neutralizer at an appropriate concentration.

Concentrations of phenolic disinfectants

As conditions in discard jars vary from one laboratory to another, formulations of proprietary disinfectants change and new products appear on the market, it is not possible to make firm recommendations for concentrations of phenolic disinfectants. Those who are in charge of a laboratory must make themselves responsible for this decision, following the results of in-use testing. As a starting point, it is suggested that the highest concentration recommended by the manufacturer, or in due course, by the Department of Health and Social Security, be used in the discard jar. Effectiveness may be checked by the in-use test and the concentration increased if there are signs of disinfection failure, once it is certain that the disinfectant solution is being made up correctly as described below (management of discard jars).

Concentrations of hypochlorite disinfectants

Proprietary hypochlorite solutions contain different amounts of available chlorine which is usually, but not always, stated in the literature. These are normally 100,000, 50,000 and 1000 parts per million (ppm) of available chlorine. Users are warned that one brand is sold in two concentrations without this being specified in the advertising literature. If there is a doubt concerning the concentration of the solution purchased, it may be assessed chemically in the laboratory. We recommend a solution containing 100,000 ppm of available chlorine, to be used in the discard jar at a concentration of 1·0% (1000 ppm available chlorine). In the presence of gross soiling with human blood, a concentration of 10·0% (10,000 ppm available chlorine) is recommended in the discard jar, because of the risk of infection with hepatitis virus and the fact that the solution is severely inactivated by the blood. Starch-potassium iodide paper obtainable from Johnsons of Hendon may be a useful on the spot check for the presence of avialable chlorine remaining in a solution. One such paper dipped in the disinfectant will show a deep purple-black colour in the presence of at least 200 ppm available chlorine, a concentration which is usually considered adequate for disinfection. A colour on the starch-potassium iodide paper which would be considered too pale for writing ink, means that too little chlorine remains for disinfection purposes. One of the sodium hypochlorite solutions on the market is coloured pink and the colour gradually disappears in the presence of organic material but the pink colour is not intended to be used as an indicator of chlorine content. It is added by the manufacturers to distinguish their product from other hypochlorite solutions many of which are yellow.

The suggested concentrations for hypochlorite disinfectants which have been given are named as starting points only. As with phenolic disinfectants, in-use testing must be carried out as a guide to a firm decision on a final concentration to be recommended for use, and subsequently as a method of routine monitoring. The starch-potassium iodide check method is not intended as an alternative to in-use testing, but it may be a useful guide when in doubt as to whether sufficient chlorine for disinfection continues to be available in one particular discard jar.

Management of Discard Jars

The management of discard jars generally receives too little attention in a laboratory. If neglected and mis-used, the disinfectant in the jar may not only fail to destroy microorganisms but may in fact provide a medium for their growth. A number of outbreaks of infection in hospitals have been traced to contaminated disinfectants (Bassett, 1971). In dilution, many of them gradually lose their effectiveness. If a dilution is prepared some days before it is put into use or if it is used at once after preparation but kept in use for several days, then some organisms may not only survive but grow. Maurer (1969) has given an account of the increase in numbers of pseudomonads in some disinfectant solutions which have slowly lost effectiveness after dilution. An example of *Pseudomonas aeruginosa* NCTC 6749 growing in a phenolic disinfectant is shown in Fig. 2. This shows that 24 h after inoculation of a freshly prepared dilution of the disinfectant, no surviving organisms could be detected in a small sample. After one week, the very small number of survivors which must have been present had multiplied to give more than 10^6 organisms/ml disinfectant. The disinfectant dilution which was prepared 7 days before inoculation had lost some effectiveness so that 10^2 survivors/ml were present after 24 h and they multiplied rapidly during the next 6 days. The same effect may be shown in disinfectant solutions which are incorrectly measured and too weak to kill even when freshly prepared, or when the solution is inactivated by organic or other material.

Correct measurement of the disinfectant concentration is all important. At a convenient height on the outside of every discard jar a line should be drawn with a felt pen to show the level of a known volume of solution. The clean jar is approximately half filled with water. The correct volume of disinfectant is added to water and more water is added to bring the level up to the felt pen line, measured by pipette, or in a small marked beaker or marked universal container which is kept for the purpose.

The disinfectant solution prepared in this way should be kept in use for one day only. Pipettes discarded into it should be left overnight and next morning the jar should be emptied, washed and disinfected by heat (65°

/10 min) to destroy any organisms which may be surviving in the jar. If such organisms are present, they are likely to have acquired some resistance to the disinfectant and if carried over into the fresh solution may survive and multiply. Disinfection by heat is more reliable than a chemical method

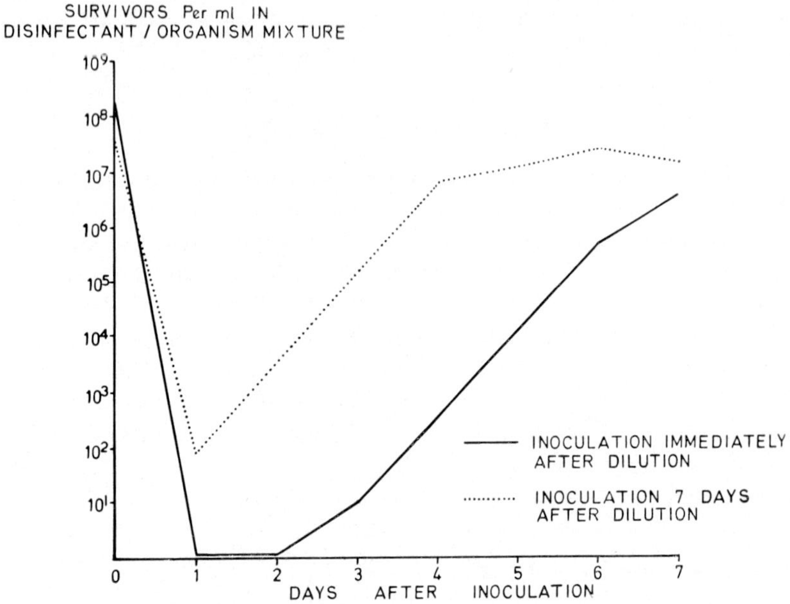

FIG. 2. Survival and growth of bacteria in a disinfectant. Reproduced by permission of the Editor of the *Pharmaceutical Journal*.

(see also p. 18). The clean disinfected jar is used for the preparation of a fresh solution.

This routine should be followed on all occasions even when the disinfectant in the jar has not been used at all but has been standing for 24 h after preparation. It may appear extravagant to throw away a jar full of unused solution but this is preferable to risking a disinfection failure as a result of deterioration of the disinfectant after dilution.

The in-use test is invaluable in detecting failures in disinfection which may be due as much to lack of care in management as to an unsuitable choice of disinfectant or selection of a concentration which is too low. In cases of disinfection failure it is suggested that:

1. The method of making up the contents of the discard jar be checked while watching for accurate measurement in a clean jar which is disinfected by heat and the solution renewed daily; "topping up" the disinfectant in the jar is bad practice.

2. The concentration of the disinfectant be increased and the addition of a compatible detergent be considered if the load of organic material is heavy.

3. Continuing failure may indicate that the task is beyond any chemical disinfectant, in which case heat treatment is suggested.

No chemical disinfectant can be depended upon to sterilize, but if used with care and regularly checked by in-use tests, the products suggested here will normally give safety in discard jars.

References

ANON (1960). British Standard 3286: *Method for laboratory evaluation of disinfectant activity of quaternary ammonium compounds*. London: B.S.I.

BASSETT, D. C. J. (1971). Common source outbreaks. *Proc. R. Soc. Med.*, **64,** 980.

KELSEY, J. C. & MAURER, I. M. (1966). An in-use test for hospital disinfectants. *Mon. Bull. Minst. Hlth.*, **25,** 180.

MAURER, I. M. (1969). A test for stability and long term effectiveness in disinfectants. *Pharm. J.*, **203,** 529.

RUBBO, S. D. & GARDNER, J. F. (1965). *A Review of Sterilisation and Disinfection*. p. 130. London: Lloyd-Luke.

Notes on Hot Air Sterilization

R. ELSWORTH

*New Brunswick Scientific Co., Inc., New Brunswick, New Jersey 08903, U.S.A.**

The facts of hot-air sterilization were established by Bourdillon, Lidwell and Lovelock (1948). Thus it became accepted at that time that to sterilize dry, spore-laden air an exposure time of 1 sec at a temperature of 300° was a satisfactory combination. The laboratory tests to support this contention showed that the survival was not greater than 1 in 4000. However, in using this design value to construct units which supplied air to deep culture vessels it was clear that the actual reduction achieved was much greater and that the survival was not more than 1 in 10^8 (Elsworth, Telling and Ford, 1955).

Subsequent work has done no more than refine the original pronouncement of Bourdillon *et al.* (1948). First Gaden and Humphrey (1956) estimated the minimum efficiency which is required from a filter or hot air sterilizer in order that it shall not produce a contaminated culture. They deduced that air supplied to a typical mould fermentation should have its original spore count reduced by a factor of 1 in 10^{15} ($10^{-13}\%$). Then Elsworth, Morris and East (1961), noting the discrepancy between this value (1 in 10^{15}) and their estimate of the ultimate sensitivity of a practical laboratory test (1 in 10^9), proposed a method of extrapolation by which it was possible to **estimate**, from laboratory data, values at this lower level of survival. All extrapolations depend on faith, and the ultimate test is whether units designed on this basis work in practice. In fact, as in the case of those based on Bourdillon's results, they do.

These notes, in summary form, are concerned with the following:

1. The minimum removal efficiency necessary in various circumstances.
2. Theory of heat sterilization.
3. An extrapolation method of testing a sterilizer.
4. The relation between temperature *versus* exposure time to effect a spore reduction of $10^{-13}\%$ (1 in 10^{15}). This is needed to design a sterilizer.

* Address for correspondence: 25 Potters Way, Laverstock, Salisbury, Wiltshire, England.

The Required Removal Efficiency

Elsworth (1969) calculated that if the chance of a contaminated culture is to be not more frequent than 1 in 1000, then for a 24 h culture using 200 ft^3/min of air the sterilizer must reduce the spore count of the supply air by a factor of 10^{-7}%. In a smaller scale culture using only 0·1 ft^3/min the reduction need only be 10^{-4}%. In order to sterilize effluent air (which contains an excess of microorganisms) the allowable penetration is only 10^{-11}. The temperature coefficient of penetration is high; there is a 10-fold reduction for approximately 10° rise in temperature and, conversely, a 10-fold increase for a 10° fall in temperature. Consequently to allow for crudity in temperature control, it is wise to set a design value for all purposes of 10^{-13}% penetration. In so doing, the power consumption for heating will be 10–20% greater than in the case when a 10^{-4}% penetration would suffice.

Theory of Heat Sterilization

Experimentally the death of microorganisms whether by dry or moist heat fits the familiar exponential equation (see p. 70 for symbols):

$$1/N_t \cdot dN_t/dt = k_T \qquad (1)$$

where N_t is the number of organisms surviving at time t and k_T is a velocity constant.

Rahn (1945) deduced that hot air sterilization of spores is an oxidation reaction which will obey the Arrhenius equation:

$$\log_{10} k_T = A_1 - \mu/(2{\cdot}3\ RT) \qquad (2)$$

where A_1 is a constant, μ is the activation energy, R is the gas constant, and T is temperature (°K). He deduced that μ should have a value around 10,000 cal/mol. Thus by plotting $\log_{10} k_T$ against $1/T$, the data should fit a straight line the slope of which will be: $-\mu/2{\cdot}3\ R$.

The value of k_T may be found by measuring N_0, the entrant concentration of spores in the air stream, and N_t the concentration of survivors in the effluent air, and substituting in the equation:

$$k_T = \frac{(\log_{10} N_0 - \log_{10} N_t) \times 2{\cdot}303}{t} \qquad (3)$$

$$= \frac{(2 - \log_{10} P) \times 2{\cdot}303}{t} \qquad (3a)$$

where P is percentage penetration (100 N_t/N_o). However, because N_t is so small it is impossible to determine with sufficient accuracy its value at the operating temperature corresponding to virtual sterility. By working over a range of temperature below this value, where N_t is a measurable quantity, k_T can then be calculated from equation (3). As stated above, the plot of k_T *versus* $1/T$ should lie on a straight line. The value of k_T which corresponds to the desired penetration can be calculated from equation (3). The corresponding temperature can then be found by extrapolating the experimental plot of k_T versus $1/T$ (equation 2). This forms the basis of the method of testing.

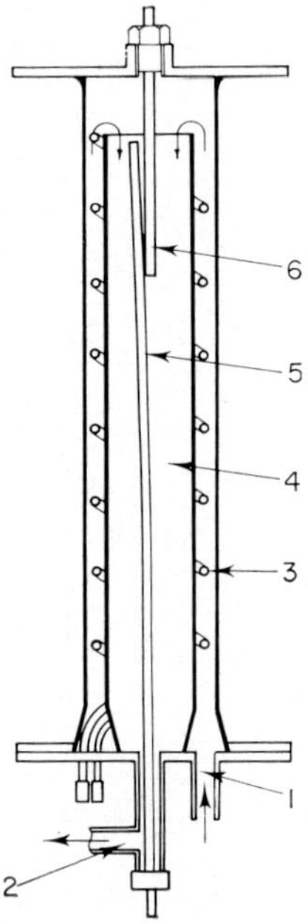

FIG. 1. A 100 l/min air sterilizer; 1, inlet air; 2, exit air; 3, heating element (1 kW max); 4, exposure chamber; 5, thermocouple in sheath (used during testing for measuring temperature gradient), and 6, mineral insulated thermocouple used as detector for temperature controller.

Examples of heat sterilizers

Figure 1 shows the unit described by Elsworth *et al.* (1961). It is made from a 2 ft length of 3 in. pipe. The exterior is lagged and the internal heating element (1 kW) is operated by an on-off controller. The unit is designed to operate at 300–350° when its capacity is 100 l/min of free air. Heat regeneration is not provided.

A proprietary unit made by the New Brunswick Scientific Co., Inc. is shown in Fig. 2. The sterilizing unit is $6\frac{5}{8} \times 20$–21 in. There is a heat exchanger between the inlet and exit air, followed by a water cooler. The heater is 2·8 kW and operates through an indicator-controller. The control

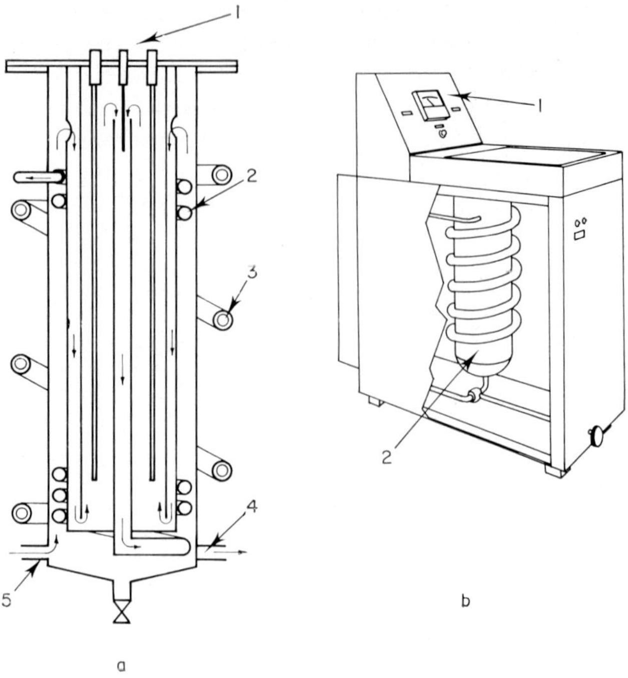

FIG. 2. The NBS air incinerator (Series CN–10).
(a) **The exposure chamber** (the arrows show the direction of the air flow); 1, entry points for 4 heaters and pyrometer; 2, heat exchanger; 3, water cooler; 4, air outlet, and 5, air inlet. (b) **Exploded view of complete console** (the cabinet measures $14\frac{1}{2} \times 31$ in. at the base, is 42 in. high and a single unit weighs 200 lb net.); 1, control panel and temperature indicator, and 2, exposure chamber.

system is built into the console and includes safety devices such as a current limiter, a second thermostat, and an air shut-off if there is power, water or heater failure. The unit is designed to run at a maximum temperature of 370°. The rated capacity is 200 l/min of free air.

A much bigger unit, designed to treat 550 ft^3/min of air from animal exposure chambers, has been described by Chatigny, Sarshad and Pike (1970). A natural gas burner heats the air to 400°. To quote from the paper: "Tests showed a reduction in number of aerosolized viable hardy spores . . . of more than 8 logs at design flow rates". This is probably a conservative estimate.

Testing a Heat Sterilizer

Sensitivity of the test

Fresh air, at least in our experiences in Wiltshire, rarely contains more than

3 bacteria-bearing particles/ft^3 and compressed air contains less. At such a concentration, to **measure** a penetration of $10^{-10}\%$ would require an impossibly large air sample of 10^{12} l. Increased sensitivity will result if the bacterial concentration in the air supply is increased. This is arranged in a manner analogous to that used in filter testing. A spore suspension is atomized and the aerosol created injected into the air stream which supplies the air sterilizer. One atomizer of the type described will give a maximum concentration of 10^6 spore/l in an air stream of 100 l/min. For larger air streams a bank of atomizers must be used.

The largest air sample it is practicable to take rarely exceeds 1500 l. If no surviving organisms are detected there is still a 20-1 (5 %) chance that there may be as many as 3 organisms in the sample (Elsworth *et al.*, 1955). Thus the limit of sensitivity of the test is given by:

$$\frac{100\ N_t}{N_o} = \frac{100 \times 3}{1500 \times 10^6} = 2 \times 10^{-7}\%$$

Apparatus

A suitable aerosol generating apparatus described by Elsworth *et al.* (1961) is shown in Fig. 3. It is similar to that used in the methylene blue (Anon,

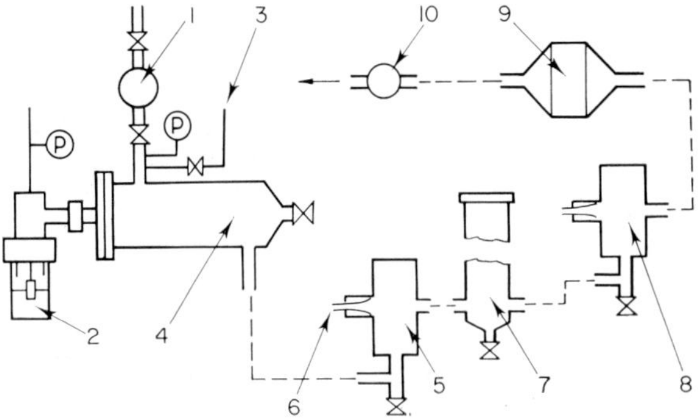

FIG. 3. Test equipment; 1, rotameter; 2, atomizer; 3, steam supply; 4, evaporation chamber; 5, sampling chamber; 6, inlet sample point; 7, sterilizer under test; 8, exit sample point; 9, bacterial filter; 10, rotameter, and P, pressure gauge.

1955) or the Sodium chloride particulate (Anon, 1969) tests for respirator filters. The atomizer (2), which is described elsewhere (Anon, 1955), discharges into an evaporating chamber (4) designed to give a holding time of 5–10 sec. The air, now containing a monodispersion of dry spores, then passes through a sampling chamber (5) the sample outlet of which is tapered from 1 to $\frac{1}{4}$ in. diam. Iso-kinetic sampling is not necessary. Next comes the sterilizer under test (7), followed by another sampling chamber

(8). The air leaves through a bacterial filter (9) and a rotameter (10). The filter serves to trap spores when tests are being made at temperatures at which there is appreciable penetration.

Operation

Before use the apparatus is disinfected. The atomizer is replaced by a nipple fitted with a purge cock, and likewise the filter (6) and the rotameter (7). The equipment is then treated with steam through branch (3) (20 psig) for 30 min. After steaming, the filter and rotameter are replaced and the air rate adjusted to its working value. The sterilizer is then raised to its working temperature. The spray bottle (1) is re-connected after charging with 200 ml of a pasteurized spore suspension to *c.* 1 in. below the jet. The concentration of the suspension varies from 10^4–10^{10} spores/ml according to the particle count which is required in the air stream. One charge permits of a minimum of 45 min spraying before the contents need to be topped-up with a further 10 ml of suspension. By adjusting the air supply the pressure drop across the atomizer is set at 30 psig. Air consumption is about 7 l/min. The main air is then reduced to compensate for the additional supply from the atomizer.

Air sampling

Before testing at higher temperatures, the apparatus is operated at room temperature to discover whether there is deposition of spores by impaction between the two sample points. If there is, then results should be corrected for this effect. The equipment shown in Fig. 3 does not have any impaction losses. However, this factor should always be tested in case the sterilizer itself acts as an impactor.

Tests at high values of penetration

A micro-filter used as described in detail by Elsworth *et al.* (1955) is a convenient sampler. It is sterilized before use and then handled aseptically. The filter bed ($\frac{1}{2} \times \frac{5}{8}$ in.) is packed with 0·16 g Merino wool-asbestos mixture. The collection efficiency for 1 μ particles, at the sampling rate of 10 l/min is $>99\%$. The air test is aspirated through a filter for a measured time. A critical orifice in the suction line controls flow. The filter wads are transferred to 1 oz screw capped bottles containing 10 ml of a sterile aqueous solution of 0·05 (w/v %) Manucol (Alginate Industries Ltd.). The bottles are shaken to disperse the entrained spores and colony counts are then made on the supernate.

If, for example, the inlet aerosol gives 100 colony counts/l, and a 15 min sample is taken at 10 l/min, then the count of the filter extract will be 1500 colonies/ml. The sample of exit air should be taken for a longer period according to the degree of sterilization.

Tests at minimum values of penetration

To test at higher sensitivity ($P \sim 10^{-7}\%$) the concentration of spores in the inlet air is raised to $\sim 10^6$ colony counts/l by using a stronger suspension in the spray bottle. A micro-filter is used to examine the inlet air but a sample of only one minute's duration need be taken.

A large sample of exit air may be taken by using a slit sampler (Bourdillon *et al.*, 1948). If samples are taken on four successive plates over a period approaching one hour, the sample size will be *c.* 1700 l. For a typical calculation see under "Sensitivity of thc test" above.

Reaction time

This depends on the volume of the heating and holding chambers, and the temperature profile—details given by Elsworth *et al.* (1961).

Typical results

The results considered are taken from those of Elsworth *et al.* (1961) and given in abridged form in Table 1. They are used to deduce: (a) the order of the reaction, (b) the expected penetration at a working temperature of 330°, and (c) to construct a graph of the temperature-time relation to effect a penetration of $10^{-13}\%$. This can serve as the basis of design of other units.

The order of reaction

Figure 4 is a plot of the values of $1/T$ versus $\log_{10}k_T$ taken from Table 1.

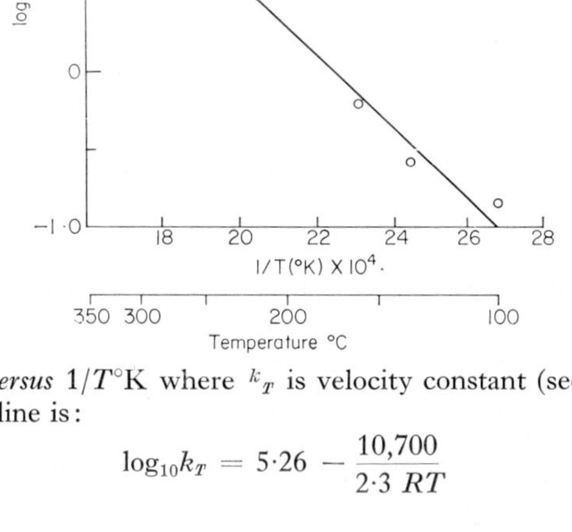

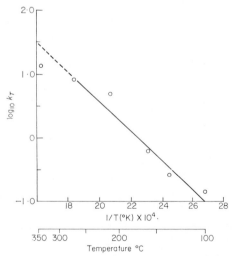

FIG. 4. $\text{Log}_{10}k_T$ *versus* $1/T$°K where k_T is velocity constant (sec^{-1}); the equation of the best fitting line is:

$$\log_{10}k_T = 5{\cdot}26 - \frac{10{,}700}{2{\cdot}3\ RT}$$

TABLE 1. Heat destruction of spores—experimental results*

Reaction temp (°C)	Penetration (%)	Reaction time (sec)	Velocity constant, ^{k}T (sec^{-1})	$\log_{10}{}^{k}T$
100	78	1·77	0·143	−0·845
135	66	1·62	0·256	−0·592
160	38	1·54	0·630	−0·201
210	0·1	1·40	4·94	+0·694
270	$3{\cdot}6 \times 10^{-3}$	1·26	8·12	+0·910
330	$<1{\cdot}75 \times 10^{-7}$	1·15	>17·5	> +1·24

* Elsworth *et al.* (1961).

The calculated equation of the line of best fit has a value of $\mu = 10{,}700$ (equation 2). The linearity and the low value of μ support Rahn's view that the reaction is first order and that the process corresponds to a chemical oxidation.

Penetration at 330°

By extrapolating the line of best-fit in Fig. 4 the values of k_T at higher temperatures can be deduced. Using equation (3), these can be transformed into penetration values. The results plotted in Fig. 5. show the high tem-

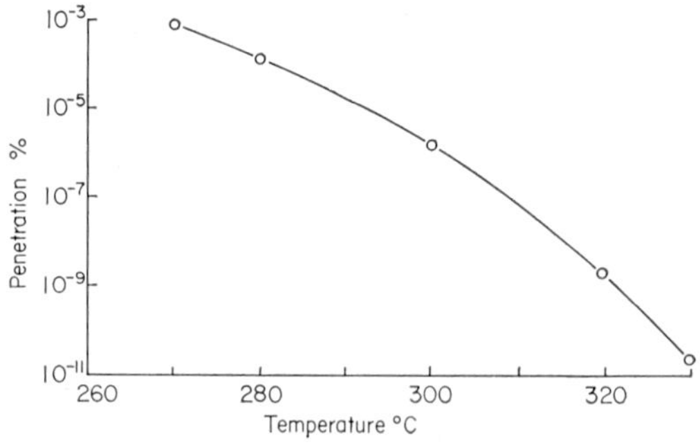

FIG. 5. Calculated penetration values obtained by extrapolation of the best fitting line in Fig. 4.

perature coefficient of penetration. This underlines the difficulty in reproducing results within and between various units, unless there is precise temperature measurement and control.

The value of penetration at 330° which lies between 10^{-10} and 10^{-11}% is put forward as a better estimate than the experimentally determined value of $<1{\cdot}75 \times 10^{-7}$% given in Table 1.

Relation between Temperature and Exposure Time

(Recommended for use in designing a heat sterilizer)

The equation of the best fitting line in Fig. 4 is of general application to spore destruction:

$$\log_{10} k_T = 5{\cdot}26 - \frac{10{,}700}{2{\cdot}3\ RT} \tag{4}$$

Given that the desirable penetration (P) is $10^{-13}\%$, then substituting this value in equation (3a) gives the relation between k_T, the velocity constant, and the exposure time, t:

$$k_T = \frac{(2+13)\times 2{\cdot}3}{t} = \frac{34{\cdot}4}{t} \tag{5}$$

By combining equation (4) and (5), t can be expressed as a function of $1/T$, given by:

$$\log_{10} t = \left[2320 \times \frac{1}{T}\right] - 3{\cdot}72 \tag{6}$$

after substituting 2 cal/mol for R.

Equation (6) like equation (4) applies to spore destruction processes in general.

Solving equation (6) for $T = 200$–$300°$ we get

Temperature		Exposure time
°C	$1/T$°K	t (sec)
200	$21{\cdot}1 \times 10^{-4}$	15·1
250	$19{\cdot}1 \times 10^{-4}$	5·1
300	$17{\cdot}5 \times 10^{-4}$	2·1
350	$16{\cdot}1 \times 10^{-4}$	1·05

This relation is shown graphically in Fig. 6.

If a different standard of penetration is required equation (6) may be recalculated. For instance if $P = 10^{-4}\%$, then it will be found that at 350° an exposure time of 0·76 sec will suffice compared to 1·05 sec above.

Conclusion

These notes show that heat sterilization is a reliable as well as a technically feasible process. Heat sterilizers were developed when filtration of bacterial particles was improperly understood and filters were not always reliable. Development in filtration theory and the advent of improved filter materials has now made filters technically competitive with heat treatment. Compared to heat treatment the running cost of filtration is lower and the equipment less sophisticated. Nevertheless in some circumstances, such as the disinfection of pathogenic viral cultures, there is a case for using heat sterilization because, in spite of higher cost, it is of marginally greater reliability.

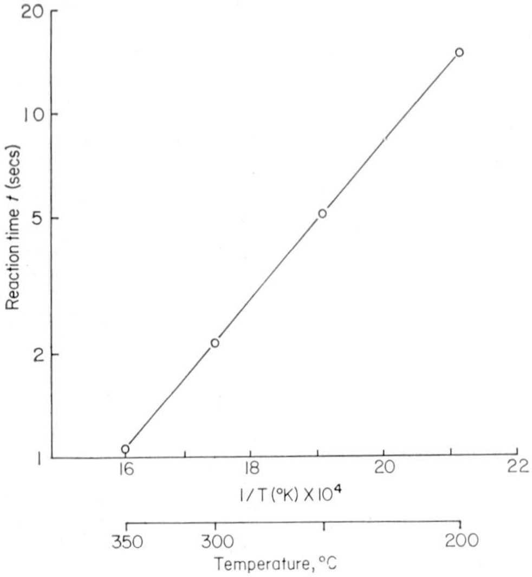

FIG. 6. Data for designing a heat sterilizer; the plot of reaction time (t) versus $1/T$ to give a penetration value for spores of 10^{-13}%; the equation to the line is:

$$\log_{10} t = \left[2320 \times \frac{1}{T}\right] - 3{\cdot}72$$

Symbols

A_1, constant in Arrhenius' equation
k_T, velocity constant at temperature T°K (sec^{-1})
N_o, inlet concentration of viable bacterial particles in the air stream
N_t, exit concentration of viable bacterial particles in the air stream
P, % penetration (or survival) given by (100 N_t/N_o)
R, gas constant, (2 cal/mol °C)
T, reaction temperature, (°K)
t, reaction time (sec)
μ, activation energy of sterilization process, (cal/mol)

References

ANON (1955). Methylene blue particulate test for respirator canisters. *B.S.* **2577** (1955). London: B.S.I.

ANON (1969). Sodium chloride particulate test for respirator filters. *B.S.*4400 (1969). London: B.S.I.

BOURDILLON, R. B., LIDWELL, O. M. & LOVELOCK, J. E. (1948). *Studies in Air Hygiene*. London: H.M.S.O.

CHATIGNY, M. A., SARSHAD, A. A. & PIKE, G. F. (1970). Design and evaluation of a system for thermal decontamination of process air. *Biotechnol. Bioengng.*, **12,** 483.

ELSWORTH, R. (1969). Treatment of process air for deep culture in *Methods in Microbiology* (J. R. Norris and D. W. Ribbons, eds) Vol. 1, p. 123. London and New York: Academic Press.

ELSWORTH, R., TELLING, R. C. & FORD, J. W. S. (1955). Sterilization of air by heat. *J. Hyg., Camb.*, **53,** 445.

ELSWORTH, R., MORRIS, E. J. & EAST, D. N. (1961). The heat sterilization of spore infected air. *Trans. Instn. chem. Engrs.*, **39,** A47.

GADEN, E. L. JR. & HUMPHREY, A. E. (1956). Fibrous filters for air sterilization, design procedure. *Ind. Engng. Chem. ind.*, **48,** 2172.

RAHN, O. (1945). Physical methods of sterilization of microorganisms. *Bact. Rev.*, **9,** 1.

The Routine Control of Contamination in a Culture Collection

I. J. BOUSFIELD AND A. R. MACKENZIE
National Collection of Industrial Bacteria, Torry Research Station, Ministry of Agriculture, Fisheries and Food, Aberdeen, Scotland

The control of contamination in most microbiological laboratories is not normally a major problem, unless the techniques of individual workers are particularly bad. The microbiologist usually "knows" his cultures and is able to detect readily the presence of unwanted organisms, in which case either he purifies the suspect culture or he discards it and obtains a replacement. If any of his cultures do become contaminated, the problem is essentially a private one in that the work of other laboratories is unlikely to suffer because of it. Also, since he will normally repeat any experiments involving contaminated cultures before publishing his results, his reputation for reliability is unlikely to be damaged outside his own laboratory. If the microbiologist is working with organisms unfamiliar to him, he may experience difficulty in detecting contamination, but even so, it is usually only his own work which will be affected by this.

However, in a service culture collection such as the National Collection of Industrial Bacteria the problems arising from inadequate contamination control are somewhat different Firstly, the cultures produced by such a collection can be used by hundreds of workers whose research may be seriously impeded if they are provided with impure cultures. Whilst these workers should themselves check the cultures they receive, they (rightly) expect the collection from which they have bought the cultures to have taken all reasonable precautions to ensure the absence of contamination. Secondly, a national culture collection handles far more cultures than does the average microbiological laboratory. The NCIB freeze-dries some 20,000 cultures a year, both to replenish its own stocks and as part of its freeze-drying service to other laboratories. Thus the possibility of a particular culture becoming contaminated is perhaps greater in a culture collection than in the average laboratory. Also, unlike the average microbiological laboratory, a culture collection obviously cannot check for purity every individual ampoule of the cultures it produces, but only samples of these. Therefore, the preparative routine must be sufficiently reliable for the assumption to be made that if one or two ampoules from a particular batch show no contamination, then the remainder of that batch is likely to be free from contamination. Finally, many microbiologists have long memories and once a culture collection has gained a reputation for supplying contaminated cultures, it is lost only with difficulty.

The aim of the present chapter is to describe the methods of contamination control currently in operation in the NCIB. The routine for the preparation of freeze-dried cultures will be outlined, points at which contamination is most likely to occur noted and the precautions taken to guard against this discussed. Finally, the methods of checking freeze-dried cultures for purity will be described.

Preparation of Cultures

All bacterial cultures in the NCIB are preserved by freeze-drying or by L-drying (Annear, 1958). When a culture is first received for deposition in the collection it is checked for purity by plating on suitable media and by microscopical examination of stained smears. If the culture is found to be contaminated, the depositor is normally requested to send a replacement. Once the purity of the culture has been confirmed, a suitable medium in the form of agar slopes (or, occasionally, broths) is inoculated to provide material for freeze-drying. After incubation, these cultures are checked by the microscopical examination of stained smears. If no contamination is detected, the cultures are prepared for freeze-drying. No special methods are used in the setting up of cultures for freeze-drying other than the strict observance of standard aseptic techniques.

Preparation of Freeze-Drying Base

When bacterial cells are freeze-dried it is essential that they be suspended in a suitable protective medium, otherwise there may be a drastic reduction in viability. Several of these protective media, or freeze-drying bases, are known: that used by the NCIB is "*Mist. desiccans*"* (Fry and Greaves, 1951) which is a mixture of Horse Serum, glucose and Nutrient Broth. Since this mixture cannot be autoclaved because of its serum content, great care must be taken to ensure that it is sterile before it is used. In the NCIB *Mist. desiccans* is sterilized by Seitz filtration, which can take a considerable time since the mixture contains a fine precipitate and filters only slowly, even under pressure. On rare occasions, leakage has been known to occur during filtration resulting in an unsterile filtrate. Since leakage cannot be detected during filtration it is essential that the sterility of the filtrate be checked.

* Formula: 75 ml Horse Serum No. 3 (Burroughs Wellcome), 25 ml Nutrient Broth (OXOID C.M.1), 7·5 g glucose.

Merely incubating the filtered product is **not** sufficient since certain organisms are unable to grow in undiluted *Mist. desiccans* (I. J. Bousfield, unpublished observation). In the NCIB, samples of *Mist. desiccans* are diluted with Nutrient Broth and incubated along with the undiluted mixture, which has been aseptically dispensed previously in 5 ml amounts in sterile, screw-capped bottles.

Filling of ampoules

Cultures are exposed to the greatest risk of contamination while they are being dispensed into ampoules. Of necessity, this process cannot be hurried and first class aseptic technique is thus of paramount importance. The ampoules themselves are sterilized by autoclaving after the insertion of a numbered and dated filter-paper strip and a cotton plug (Fig. 1). Bacterial

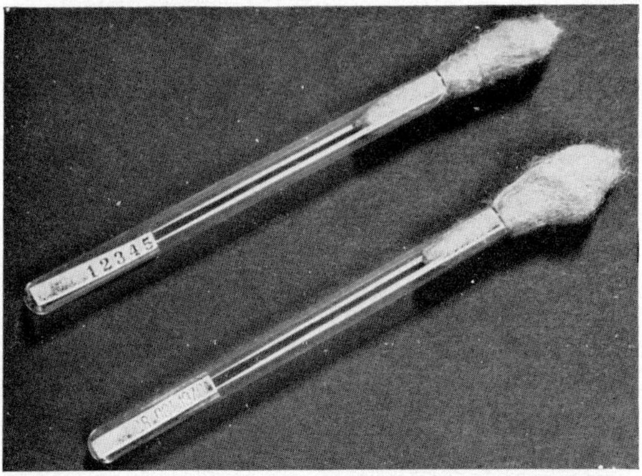

FIG. 1. Sterile ampoules containing identification slip.

growth is washed from agar slopes (or a centrifuged sediment is resuspended) with *Mist. desiccans* delivered from a sterile 2 ml pipette (Fig. 2)—the pipettes are autoclaved individually in plugged glass tubes. The resulting suspension is returned to the bottle containing the remainder of the *Mist. desiccans* thus making 5 ml bacterial suspension in all. Five drops (*c.* 0·1 ml) of this suspension are then added to each ampoule in turn, using a sterile Pasteur pipette. When the pipette has been emptied, it is discarded and a fresh one is used for filling further ampoules. When all the ampoules have been filled, the cotton plugs are replaced by sterile lint caps (Fig. 3). During freeze-drying, these caps allow water-vapour to pass through more easily than do cotton plugs. The ampoules are loaded into the centrifuge plate of the freeze-dryer and the cultures are then ready for freeze-drying.

FIG. 2. Bacterial growth washed from an agar slope with *Mist. desiccans* to give a suspension for freeze-drying.

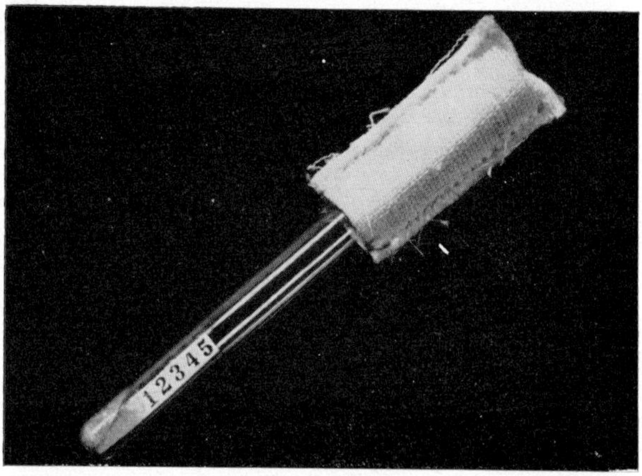

FIG. 3. Cotton plug replaced by lint cap to allow easy passage of water vapour during primary drying.

If the purity of the cultures and the sterility of the *Mist. desiccans*, ampoules and glassware has been established, there are two major possible sources of contamination: (a) the bacterial aerosol produced during the manipulation of the cultures dispensed previously, and (b) the atmosphere.

When bacterial growth is washed from an agar slope as in Fig. 2, an aerosol is produced around the mouth of the bottle (see also p. 8). When several different cultures are handled in fairly rapid succession, a mixed aerosol is built up thus providing ideal conditions for cross-contamination of cultures. The dispersal of such an aerosol depends mainly upon the movement of air around the work area, the most obvious being the upward convection currents induced by the Bunsen burner. It is generally accepted that the nearer to the Bunsen flame manipulations are carried out, the smaller the risk of contamination. However, experiments carried out in the NCIB have shown that the aerosols produced during the handling of cultures are **not** completely removed from the work area by convection currents.

Whilst sound technique can reduce the risks of aerial contamination of all kinds to a very low level, the immediate removal of airborne organisms as

FIG. 4. Vertical laminar airflow cabinet used by NCIB

soon as they appear makes the chance of such contamination very small indeed. Such conditions can be achieved using the laminar airflow cabinet shown in Fig. 4—see p. 21. Sterile (filtered) air blows continually through the cabinet from top to bottom at a speed *c.* 90 ft/min, and any organisms escaping from the operator or from the culture tubes are swept downwards through the perforated work surface. Since the air entering the cabinet is filter-sterilized, normal atmospheric contamination is eliminated. The speed of the airflow makes the formation of a persistent aerosol impossible. Thus, provided that operations are carried out with care, the risk of unwanted organisms being introduced into ampoules is negligible. The main disadvantage of the laminar airflow cabinet used by the NCIB is that a Bunsen burner cannot be used inside it. However, by modifying aseptic techniques and carrying out all "flaming" operations outside the cabinet, a Bunsen burner in the immediate working area is unnecessary.

The reduction of the contamination risk achieved by filling ampoules in a laminar flow cabinet was demonstrated by the following simple experiment. Nutrient Agar plates were arranged on the open bench as shown in Fig. 5. The tiered rack was used to "catch" extraneous organisms at various levels. After removing the lids of the plates, ampoule-filling was carried out for 2 h. The plates were then immediately removed and fresh plates were exposed for 20 min. Both sets of plates were then incubated for 5 days at 25°. The results are shown in Figs 6 and 7. The entire experiment was then

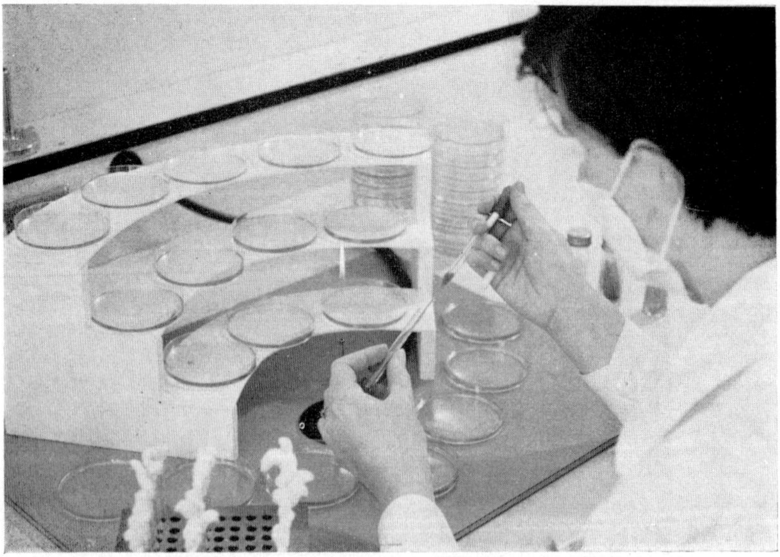

FIG. 5. Arrangement of plates to demonstrate aerial contamination.

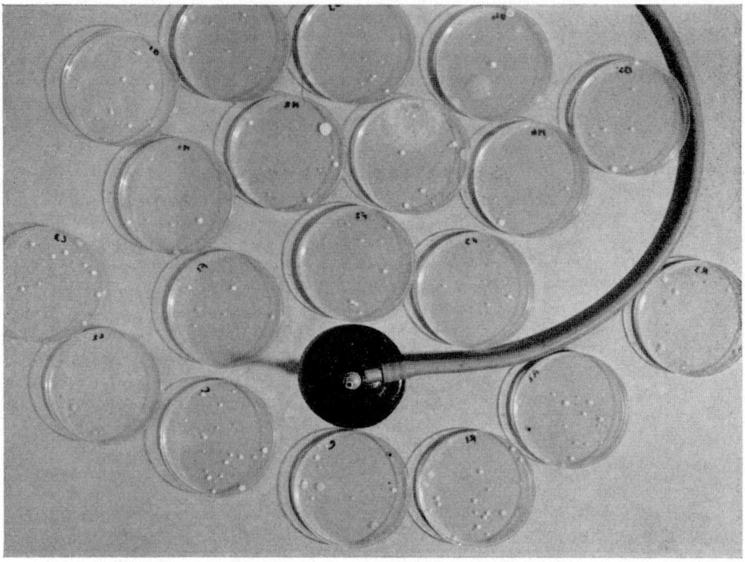

FIG. 6. Plates exposed during 2 h ampoule-filling at the open bench.

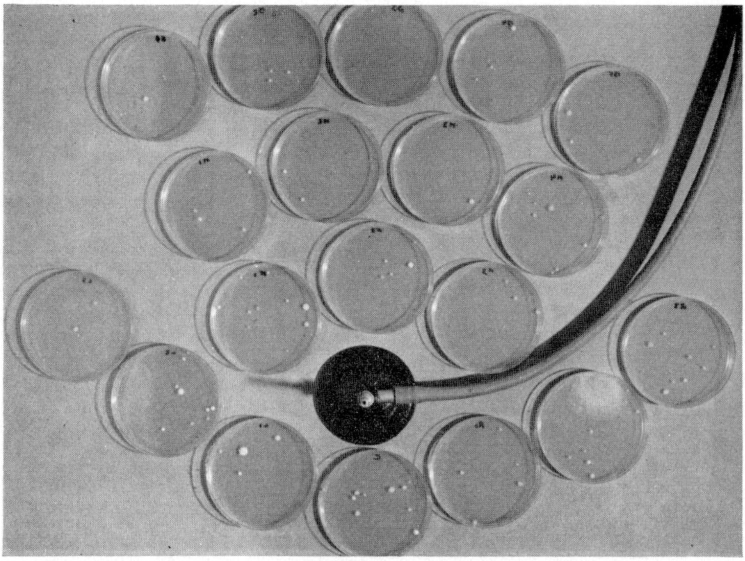

FIG. 7. Plates exposed for 20 min immediately after ampoule-filling at the open bench.

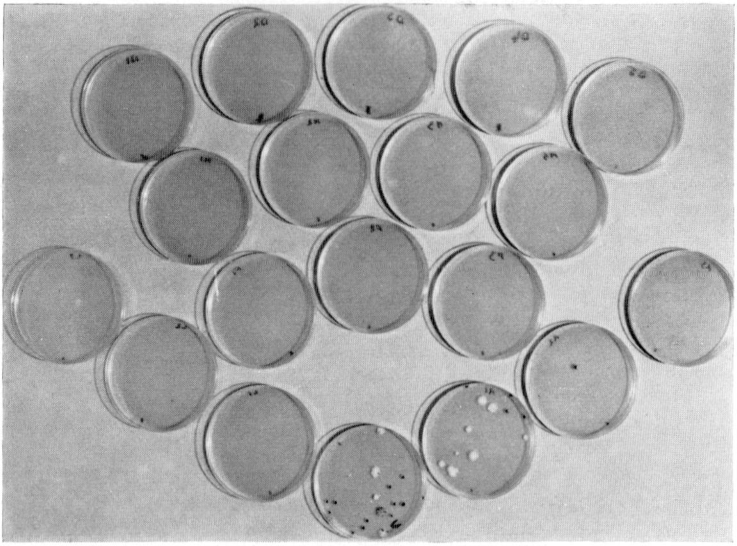

FIG. 8. Plates exposed during 2 h ampoule-filling in the laminar airflow cabinet.

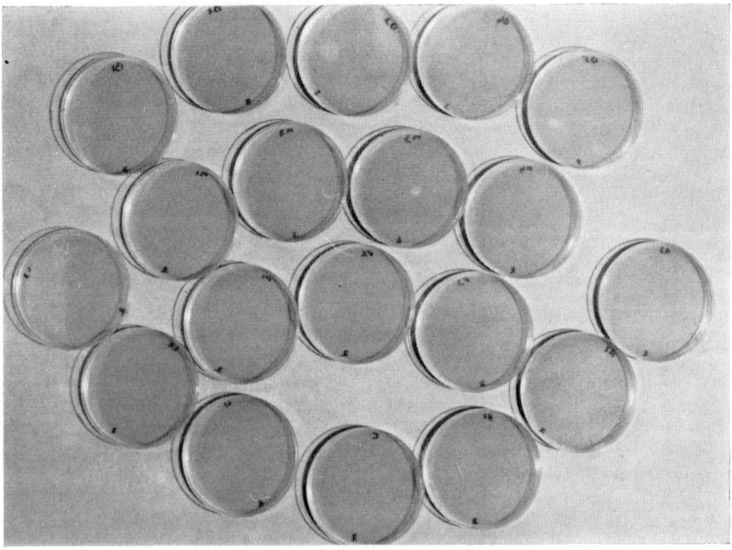

FIG. 9. Plates exposed for 2 h immediately after ampoule-filling in the laminar airflow cabinet.

repeated in the laminar airflow cabinet, the only difference being that both sets of plates were exposed for 2 h. The results are shown in Figs 8 and 9. The differences between the results of the two experiments are obvious and need no further explanation, except to point out that the three plates showing colonies in Fig. 8 were directly beneath the pipette and were contaminated solely as a result of the "fall-out" from the inoculating operations.

Freeze-drying of cultures

After cultures have been dispensed into ampoules, the chances of contamination occurring during subsequent operations are slight. Cultures are centrifuged at low speed whilst the drying chamber is being evacuated so that foaming and the resultant aerosol production are suppressed. After a few minutes the cultures are frozen solid and centrifuging may be stopped. The cultures are then left under vacuum for a further $2\frac{1}{2}$ h, during which time the bulk of their water content is removed as vapour. The cultures are then removed from the drying-chamber, the lint caps are replaced by sterile cotton plugs which are clipped and pushed down into the ampoules, which are then mechanically constricted (Figs 10 and 11). The constricted

FIG. 10. The mechanical constriction of an ampoule.

ampoules are attached to the manifolds of the secondary stage vacuum-dryer and are pumped continuously overnight to remove further water vapour, which is trapped by phosphorous pentoxide. The ampoules are then sealed off under vacuum (Fig. 12).

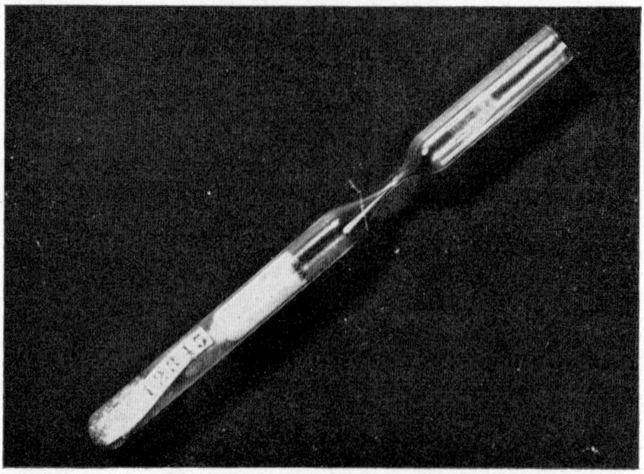

FIG. 11. A constricted ampoule ready for attachment to the manifold of the secondary stage dryer.

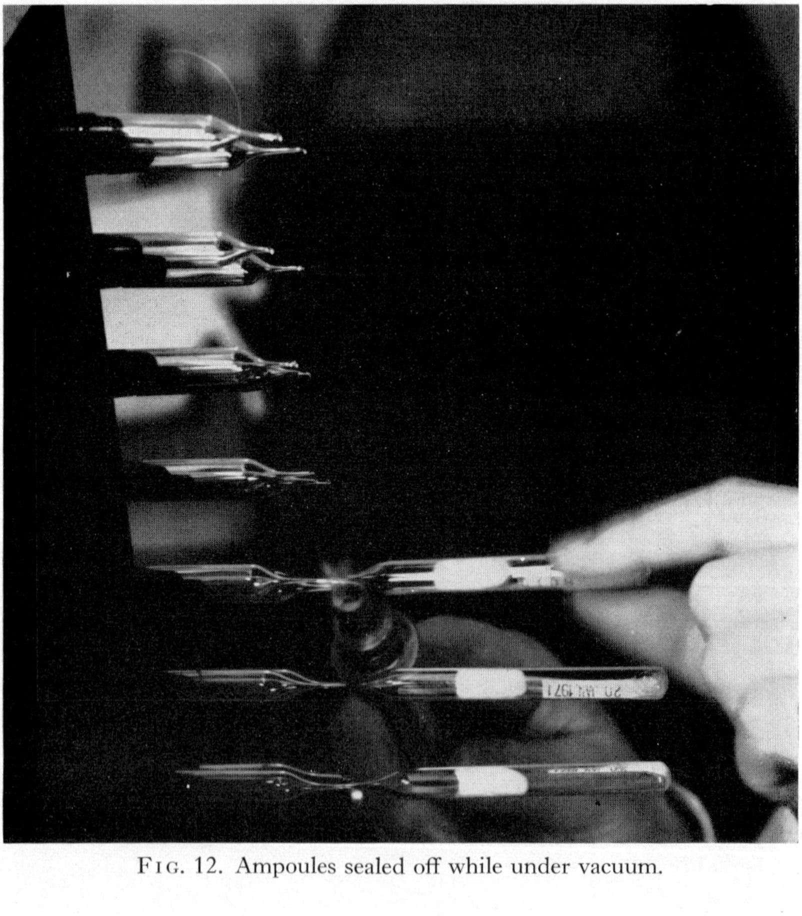

FIG. 12. Ampoules sealed off while under vacuum.

As mentioned earlier, the chances of contamination occurring during freeze-drying appear to be slight. The only point at which atmospheric contamination could be introduced is during the changeover from lint caps to cotton plugs. However, this operation is normally carried out rapidly and good technique should ensure the absence of unwanted organisms. There is a theoretical possibility of cross-contamination occurring during freeze-drying but in practice the chances are again slight. Attempts made by the NCIB to induce cross-contamination during freeze-drying by suddenly breaking the vacuum and actually blowing air into the ampoules have always failed. The lint cap over the ampoule thus appears to act as a fairly reliable filter. However, as an added precaution when cultures of more than one strain are being freeze-dried at the same time, the drying schedule is so arranged that these strains can be easily distinguished from one another. If cross-contamination should then occur, it will be easily detected. For example, cross-contamination between cultures of *Serratia, Bacillus, Micrococcus* and *Streptomyces* spp can be detected (and the cultures purified) far more easily than can cross-contamination between four strains of *Escherichia coli.*

Checking of freeze-dried cultures

When cultures have been freeze-dried, ampoules from each batch must be opened and the purity and viability of their contents checked. In a large culture collection where dozens of cultures are checked every week, the checking procedure must be as simple, yet reliable, as possible. The method described below is currently used by the NCIB as a primary checking routine.*

The entire contents of an ampoule are resuspended in a few drops of Nutrient Broth and the resulting suspension is pipetted as two separate pools on the surface of a pre-incubated, pre-dried Nutrient Agar plate. One pool of inoculum is then streaked with a wire loop so that isolated colonies will be obtained after incubation. The other pool of inoculum is left undisturbed. When the excess moisture has been absorbed by the agar, the plate is inverted in the usual way and then incubated at 25° for at least 1 week, preferably 2 weeks. After incubation, the plate is checked for extraneous colonies and Gram-stained smears are examined microscopically. If these checks show nothing untoward, then it is assumed that the culture is pure. This checking method is based on the following rationale.

1. The object of a purity check is to encourage the growth of any possible contaminants. Good growth under the test conditions of the strain being checked is immaterial. Therefore, Nutrient Agar is used and incubated for a fairly long period at 25° since it is felt that these conditions

* This method was developed by Mr. L. B. Perry of NCIB.

are likely to permit the growth of any of the "common" contaminants found in laboratory cultures. **All** cultures are plated on Nutrient Agar, even if the strain under examination will not grow, since the use of special media may inhibit the growth of possible contaminants.

2. By inoculating the plate in two places, confusion is avoided should mixed growth occur as a result of a micro-colony already present on the plate being streaked out with the inoculum, i.e. only one of the two areas of growth will show contamination.

3. As much of the growth is in the form of a confluent lawn, any contaminant colonies are easily seen against this lawn, even though the colony form of the contaminant may not differ greatly from that of the organism under examination.

4. Since the inoculum comes directly from the ampoule, the number of contaminant colonies (if any) reflects the level of contamination in the ampoule. Thus it is possible to determine roughly where the contamination arose.

5. Since part of the inoculum is streaked to give isolated colonies, if the culture is contaminated, a single colony of the desired organism can easily be picked off for purification.

6. Microscopical examination of a Gram-stained smear usually helps to confirm the presence or absence of contamination.

Figure 13 shows a typical check plate after incubation. The confluent area is even and shows no extraneous colonies. The culture is free from contamination. Figure 14 shows a single extraneous colony on the confluent area. This could be a plate contaminant, an aerial contaminant, a casual contaminant from the ampoule or the result of a very low level of contamination of the freeze-dried culture. Further ampoules must be checked. Figure 15 shows a few extraneous colonies on the confluent area. These may be aerial contaminants, although this is unlikely since they all occur on the growth patches (*cf*. Fig. 18); they may possibly be casual contaminants from the ampoule resulting from a clump of cells falling into the ampoule during filling, but this is again unlikely. There is a possibility that the *Mist. desiccans* was contaminated; if this be the case, then other cultures freeze-dried in the same batch of *Mist. desiccans* will show similar contamination. However, the most likely explanation is that one of the slopes used to grow up the culture for freeze-drying contained a contaminant colony, which became dispersed throughout the suspension used for freeze-drying. Further ampoules must be checked to confirm this. Figure 16 shows many extraneous colonies mixed with the growth of the culture being checked. Such a level of contamination would almost certainly have arisen from the freeze-drying of a mixed culture. Figure 17 shows many extraneous colonies occurring in one patch of growth only. This type of pattern suggests that

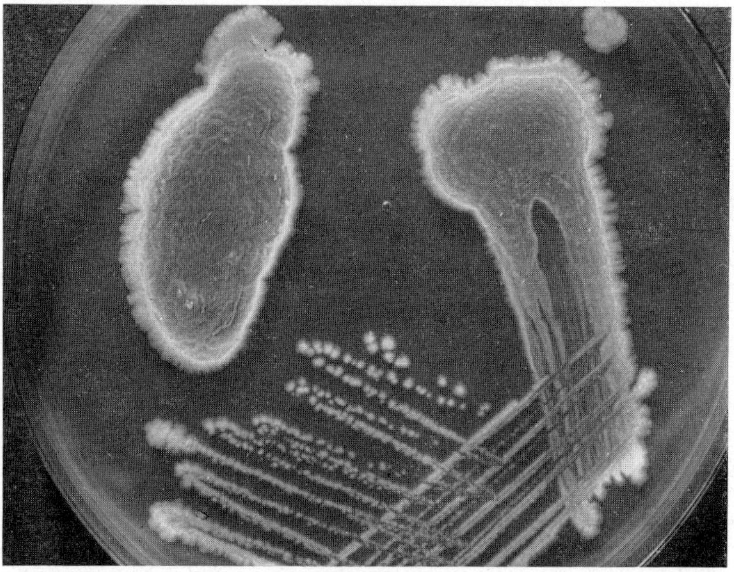

FIG. 13. Check plate showing an even confluent area with no extraneous colonies (explanation in text).

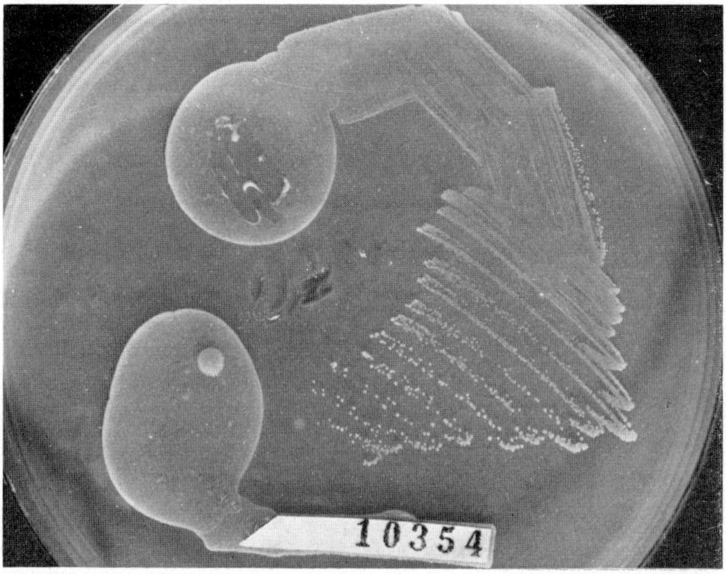

FIG. 14. Check plate showing a single extraneous colony on the confluent area (explanation in text).

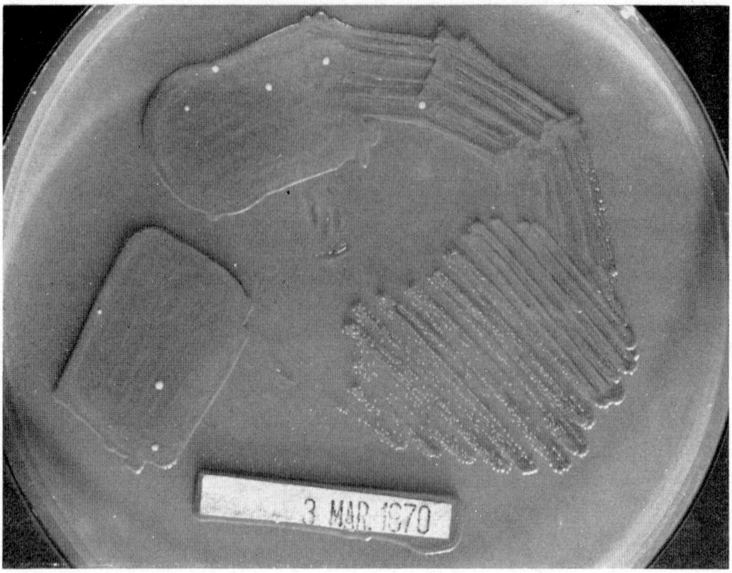

FIG. 15. Check plate showing a few extraneous colonies (explanation in text).

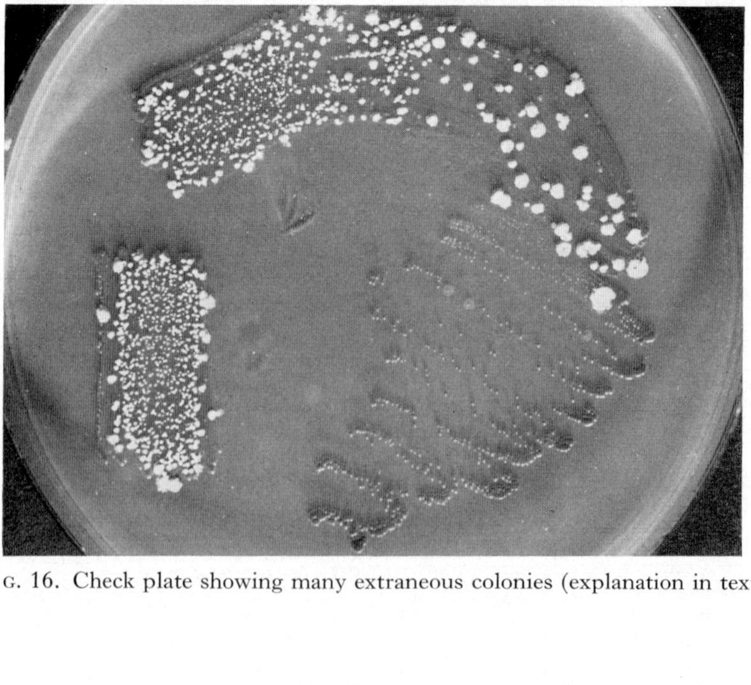

FIG. 16. Check plate showing many extraneous colonies (explanation in text).

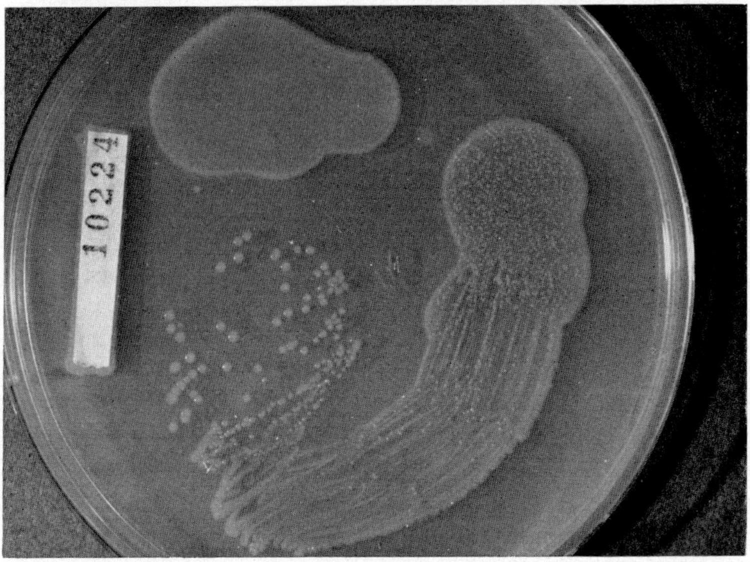

FIG. 17. Check plate showing many extraneous colonies on one patch of growth only (explanation in text).

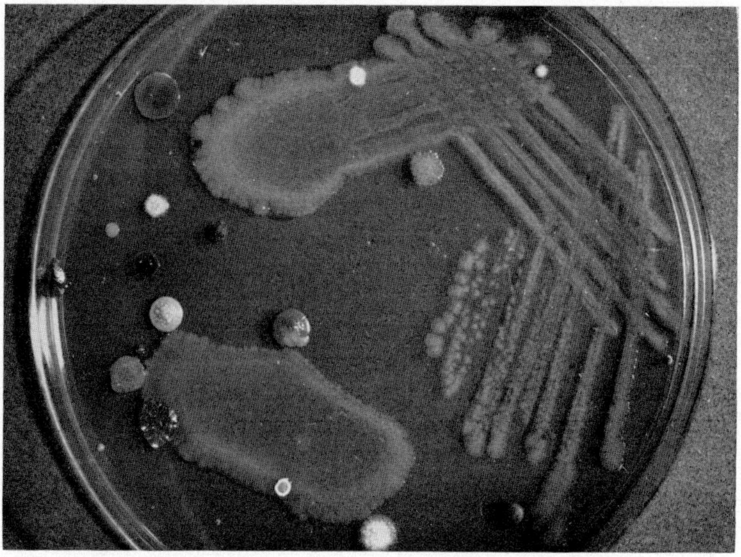

FIG. 18. Check plate showing randomly scattered extraneous colonies (explanation in text).

a micro-colony originally present on the plate has been streaked out with the inoculum. The checking of a further ampoule should confirm this. Figure 18 shows extraneous colonies scattered randomly over the plate, suggesting aerial contamination due to poor technique. The careful opening of a further ampoule should confirm this.

Once the individual worker has become familiar with this rapid checking method and the interpretation of resulting growth patterns, it seems to be very reliable and contamination is easily detected. In most cases, further checks have not been found necessary, but if there should be any doubt about the results, more intensive checks are carried out. These usually involve growing the culture under various conditions, depending upon the strain under examination. If the culture being checked does not grow under the test conditions (e.g. strict anaerobes, nutritionally exacting organisms, strict thermophiles, etc.), then obviously viability checks must be carried out separately, using appropriate techniques, after purity has been established. In addition, it may be necessary to check particular properties of the strain under examination (e.g. the requirements of nutritional mutants, assay strains, etc.). The checking routine described above is intended to detect aerobic contaminants only, since anaerobic contamination of aerobes is unlikely to be a problem. As an added precaution, anaerobic organisms are checked for contamination with other anaerobes after the absence of unwanted aerobic organisms has been verified.

Acknowledgement

This paper was prepared as part of the programme of the Torry Research Station (Crown Copyright Reserved).

References

ANNEAR, D. I. (1958). Observations on drying bacteria from the frozen state and from the liquid state. *Aust. J. exp. Biol. med. Sci.*, **36,** 211.

FRY, R. M. & GREAVES, R. I. N. (1951). The survival of bacteria during and after drying. *J. Hyg., Camb.*, **49,** 220.

Sterility Testing and Assurance in the Pharmaceutical Industry

MARGARET C. HOOPER, R. SMART, D. F. SPOONER AND G. SYKES

Quality Control Microbiology, The Boots Company Ltd., Nottingham, England

Many medicaments are administered by injection, or they are applied to the eye or to an open wound, and clearly one of the first requirements for such preparations, be they aqueous, oily or in the form of ointments or creams, is that they shall be free from contaminating microorganisms. The preparation of such a sterile product on the large scale is not easy, particularly when the nature of the product is such that it cannot be heat sterilized and aseptic manipulations have to be employed—and even when heat can be used, as with some solutions and for the sterilization of containers and closures, the reliability of the procedure in terms of temperature and time of heating must be assured.

Because the products under discussion are intentionally sterile it might be assumed that a test for sterility would give an adequate check, but this is not so. It is only possible to test a small proportion of the containers filled—the official tests in Great Britain specify 2% of the containers of each batch, and not more than 20 if the batch size is greater than 1000 containers—so that on a statistical basis alone a batch of material with a small proportion of contaminated containers could pass the official test.

The test for sterility is, in fact, only the final event of a series of events each one of which is designed to give greater assurance that proper care has been taken to eliminate the various possible sources of contamination and so that the final product is above suspicion microbiologically. These events, or steps, are: 1. the training of personnel in aseptic techniques; 2. controlling the microbiological quality of the environment in which the aseptic work is carried out; 3. checking the reliability of the equipment and the efficacy of the processes employed, and, after this, 4. the test for sterility on the finished product.

Training of Personnel in Aseptic Techniques

All persons involved in the manufacture and filling of sterile products should be given an elementary training in bacteriology, sterilization procedures and aseptic techniques. The type of instruction given varies according to the extent of the operator's previous training, and generally the whole course is completed in one week. An extension and refresher course is undertaken some months later by most of the operators.

In the first course sufficient bacteriology is taught to enable the trainee to appreciate the reasons underlying the techniques demonstrated in the course. The less previous training a person has had the more care is needed to avoid confusing him with detail. Visual aids and constant reference to hygiene in the home are useful ancillaries.

Sources of contamination are demonstrated by exposing agar plates to coughing, sneezing and talking. The effects of washing and of various skin disinfectants on the skin flora are illustrated practically. Hair and skin flakes are also cultured. Examination and discussion of these findings always proves impressive and diverting.

A series of simple exercises involving sterile filtrations and aseptic transfers, similar to those used in practice on the larger scale, is introduced to give the operator dexterity and confidence. All exercises are conducted in a sterile room prepared, under instruction, by the trainee, and as far as possible Nutrient Broth is used so that contamination can be detected easily by simple incubation. It is advantageous to have two or three people under training together to introduce a competitive aspect.

Outside the sterile room more exercises are used to demonstrate methods of sterilization such as dry heat, steam in the autoclave and heating with a bactericide, always including sublethal conditions to show where failures can occur. At the end of the course a simple written test is given to ensure that the trainee has absorbed sufficient of the facts.

In the extension course, which occupies 3 days, revisionary exercises from the first course are undertaken and further studies are made on sources of contamination, the disinfection of surfaces and other materials and on sterilization problems. Experiments are also included to familiarize the trainee with the function of the microbiologist when he makes a survey of a sterile area. The effects of clothing, sterile gowns, etc., and of movement on the shedding of organisms are investigated, and the range and limitations of the activities of various disinfectants examined. Aspects of sterilization such as the rate of penetration of heat into a load and the efficacy of gaseous methods are also considered.

Control of the Environment

Sterile rooms

The first requirement in carrying out aseptic manipulations on the large scale is for the room to be properly designed and to function correctly. This involves choosing a smooth, impervious wall finish which can be easily and regularly cleaned and disinfected, and a floor finish which is also smooth, is not subject to wear (so that dust problems are avoided), and which can withstand a daily disinfectant treatment. Suitably designed changing rooms for the staff are requisite and the room itself should be provided with air locks and a sterile air supply, with terminal filters at the point of entry of the air into the rooms. The air flow should be as directional as possible to eliminate turbulence and "dead spaces" in the room.

The reliability of the aseptic process is greatly improved by carrying out the filling and other aseptic operations within suitably designed screens or cabinets with a directional (laminar) air flow.

A daily regimen of disinfection of all floors and surfaces in the room is essential. Systematic swabbings from selected small areas of the room, as well as from the fixtures and equipment, and even from the operators' sterile gowns, keep the staff alert in this direction.

Airborne contamination

This is linked closely with the operators and their activities in the sterile area; so closely, in fact, that usually the operators are the exclusive potential source of contamination—and a continuous one. The simplest movements cause friction between the clothing and the skin, and this increases the shedding and dissemination of skin scales and organisms. Obviously the more vigorous the movement the more organisms are shed (Fig. 1). Oral

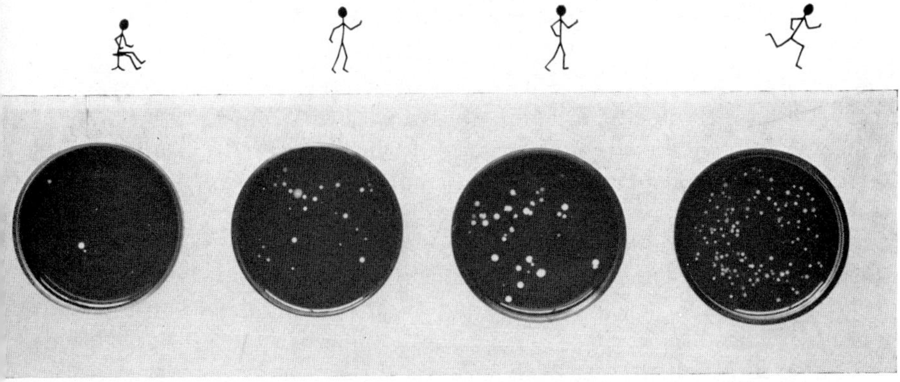

FIG. 1. The effects of movement on the relative airborne bacterial counts.

organisms are not disseminated unless the person expires violently by coughing, sneezing or otherwise creating an aerosol of saliva—normal quiet conversation releases relatively few organisms.

The extent of shedding of organisms can be controlled within limits by the type of clothing worn by the operator, an aspect which has been examined in some detail by Bethune *et al.* (1965), Bernard *et al.* (1965), Blowers and McCluskey (1965) and Sykes (1968). The effects of different types of outer garment on the level of contamination of the surrounding air are illustrated in Fig. 2 and Table 1. To obtain these, comparisons were

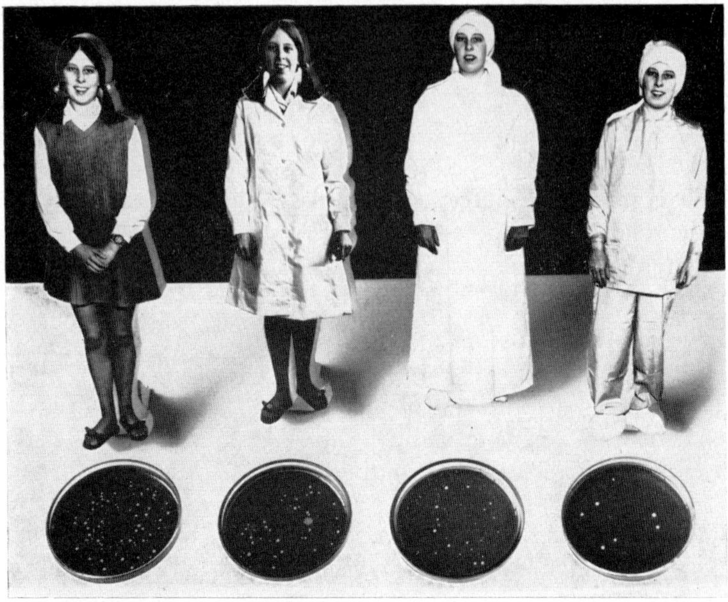

FIG. 2. The effects of outer clothing on the relative airborne bacterial counts.

made in a closed room of *c.* 850 ft^3. The operator in various types of outer-clothing first sat quietly for 3 min, then stood swinging her arms for 3 min, next walked round the room for 3 min and finally ran on the spot for 3 min. During these 4 × 3 min periods air samples were taken by means of a slit sampler and the resulting counts recorded.

Clearly the ideal clothing is that which is impermeable to airborne particles and does not allow air leaks between the clothing and the skin at such points as the neck, the wrists and the ankles, and the nearer this ideal can be attained the more effective is the clothing likely to be. But it has its disadvantages, not the least of which is the discomfort of the operator. Hence the value of using air-flushed screens and cabinets.

TABLE 1. Dissemination of skinborne organisms with different types of dress and activity

Activity	Viable bacterial count/ft³ of air with Normal dress	Laboratory coat	Sterile* gown	Sterile† trouser suit
Sitting quietly	6 (0–13)	3 (0–7)	1·5 (0–4)	0·5 (0–2)
Swinging arms	8 (6–10)	32 (7–73)	14 (1–41)	1 (0–4)
Walking	138 (20–244)	55 (25–75)	35 (4–75)	2·5 (1–4)
Running on the spot	—	140 (94–201)	53 (34–128)	4 (2–7)

* Surgical cotton gown with loose cotton trousers, overshoes, gloves and headgear.

† Nylon suit with elastic at wrists and ankles, overshoes, overlapped gloves and headgear.

Viable airborne counts

These are made with slit samplers or by exposing Nutrient Agar plates for a given period of time—for further details, see p. 37. Slit sampler counts are the more precise and informative, especially when the count is liable to vary with time. However the apparatus, complete with its vacuum pump, etc., causes much air turbulence, and so there is a preference for the settle plate method. It also has the advantage that plates can be exposed simultaneously at all levels and at a greater variety of points.

Recording and reporting results

The organisms cultured from the plates so exposed are typed to a sufficient degree to indicate their possible source—human, dust, equipment, etc. It is important to record the amount of contamination, when and where it occurred and what operation and operators were involved. A typical day's record in a sterile room, measuring 30 × 60 ft, and flushed with sterile air is illustrated in Fig. 3. In this room, maintained at a slight positive pressure, the air was introduced through a continuous line of filters situated at ceiling height along one wall of the room and extracted through similar filters near ground level on the opposite wall: the air change rate was *c.* 15/h. Points to note in Fig. 3 are:

1. the generally lower counts near (4 ft) the input filters compared with those near (4ft) the extract filters (the air at the point of entry at the filter face is sterile);

2. the broad relationship between the airborne counts and the numbers of persons in the room;

3. the relatively high airborne counts at the beginning of the day whilst the staff is arranging and settling down to the work in hand;

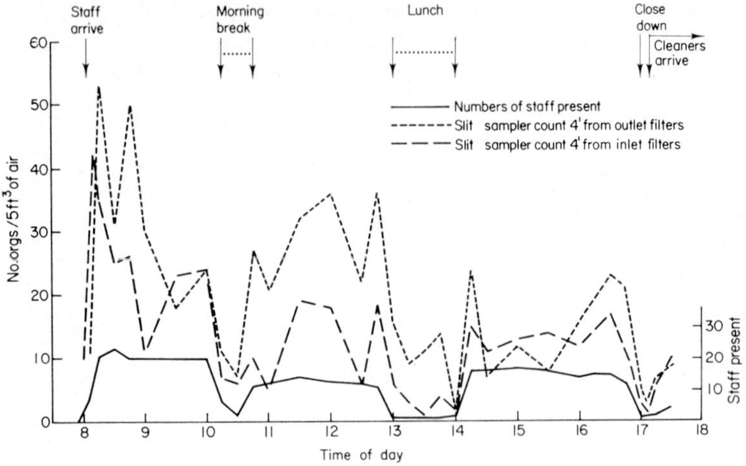

FIG. 3. Airborne counts in a sterile room in relation to the numbers and activities of staff during one day.

4. the immediate fall in count when they go off for their mid-morning break;

5. the absence of high counts on their return straight to the work in hand;

6. the fall in count during the lunch break and its return to a moderate level during the afternoon, and

7. the low counts at the end of the day before the cleaners arrive.

Sterilization Controls

Steam sterilization

Control indicators for steam sterilizations are indispensible in checking routinely against faulty packaging, loading and sterilizer performances. Their development and use have been reviewed in recent years by Sykes (1965) and Perkins (1969).

Three types are employed, biological, chemical and physical (instrumental). It is usual for the last to be a permanent feature of any sterilizer, although it is always desirable to maintain regular cross checks by means of the alternative biological and chemical controls.

Physical indicators

These are usually thermometric devices, nowadays often linked with time controllers so that not only are temperatures recorded during a sterilization cycle but also the periods for which they are held.

Siting of the heat-sensitive element is critical; traditionally the efficiency of the autoclave is checked by recording the temperature at the condense outlet at the base of the chamber. This, however, only gives a true indication of conditions within the load if air removal is adequate so that there is an uninterrupted penetration of steam. The most dependable method for checking this is by means of a heat-sensing element located in that part of the load most inaccessable to steam or to heat penetration. Sterilizers used for bottled fluids, for instance, are best monitored by means of an element placed inside one of the bottles of the load or in an exactly equivalent one permanently mounted in the chamber. It is important to ensure that faulty results are not caused by the channelling of steam along the course of the wire leads to the centre of a load. The speed of response of the temperature sensing element is also significant, especially in high speed autoclaves.

In addition to the well known types of liquid-in-glass thermometers several heat sensitive devices are used, one of the most generally acceptable being the thermocouple. This utilizes the "Seebeck effect", and the response, in the form of an electric current generated at the junction of dissimilar metals, is dependent on the temperature difference between the "hot" and the "cold" junction.

The resistance thermometer is another accurate device. In this use is made of the relationship between temperature and the variation of electrical resistance of materials, usually wires of high quality platinum or nickel. Both resistance thermometers and thermocouples require fairly extensive ancillary electrical equipment to amplify the response which can then be recorded.

Filled-system thermometers, usually mercury-in-steel, consist of a bulb, capillary tube and Bourdon spring, the last providing an essentially linear angular motion to an expressed pressure change. Mercury-in-steel thermometers are simple and self contained, but their speed of response is relatively slow. Another simple, but slow and comparatively less accurate, device is the bimetal thermometer. This utilizes a thermostatic bimetal which is a composite material made of strips of two or more metals rolled together. Because of the different expansion rates of the metals, the device changes its curvature in direct proportion to a change in temperature over a limited range.

Whatever the type of instrument it is essential to check regularly its calibration and to ensure that it is properly maintained.

Biological indicators

The use of viable organisms, particularly heat resistant spores, would seem to provide the most obvious and relevant way of controlling sterilization. However, in practice there are many difficulties. One is that, because of the inevitable delay in recovering any possible survivors by culture, they cannot be used in day-to-day control. Garden soil samples can provide a most stringent test (Spooner and Turnbull, 1942) but the species and numbers of organisms present are variable and inevitably vary in their heat resistance. Kelsey (1958) emphasized the dangers of testing the efficiency of steam sterilizers by using ill-defined spore preparations, and later (Kelsey, 1961) described the preparation of dried and standardized filter paper strips impregnated with *Bacillus stearothermophilus*; these are now available commercially (Oxoid Ltd). False positive results due to chance contamination during culture of the strips is minimized because incubation temperature is 55°, but even so care is still needed. Because they are small and thin, these test strips can be placed in critical areas in the sterilizer load.

Chemical indicators

A variety of chemical indicators has been devised. They depend on either the flow time of a fixed amount of compound having a well defined melting

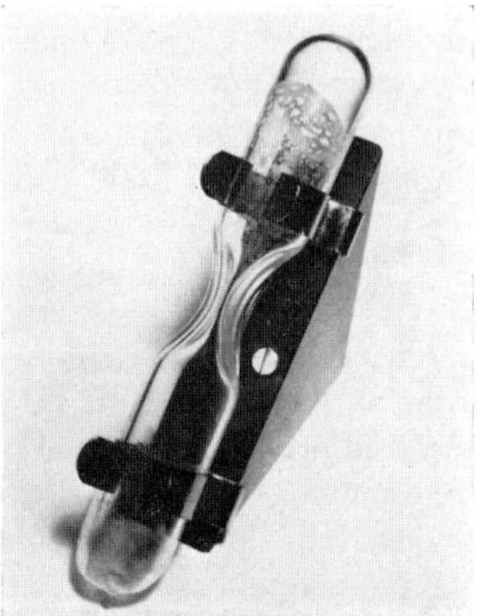

FIG. 4. The modified Brewer and McLaughlin steam sterilization indicator.

point or on a substance undergoing a time/temperature dependent reaction, e.g. the hydrolysis of an ester, which can be indicated by a colour change.

Brewer and McLaughlin (1954) utilized the flow time of a compound in an hour-glass device. A modification was introduced in these laboratories and it has withstood repeated use satisfactorily for some years (A. Royce, pers. comm.). In this modification the fragile side arm needed to eliminate air locks is replaced by the use of a simple constriction instead of a capillary (Fig. 4). The tube for control of sterilizations at 115° is made of 7/16 in. Pyrex glass tubing (wall thickness, 0·04 in.), 3–3½ in. long and containing

1 g of acetanilide B.P.C. That for 121° is $3\frac{1}{2}$–4 in. long of $\frac{5}{8}$ in. Pyrex tubing (0·05 in. wall thickness) and contains 2 g of succinic anhydride (BDH, laboratory grade). The tube is clipped at 45° to the horizontal on a wooden stand, the plane of the constriction being vertical. When the solid melts it flows through the lower portion of the slit and air is displaced through the upper part; the quantities and conditions are such that the whole of the melted solid passes through in just the required time. By varying the type, thickness and surface area of the glass the flow time can be varied to meet any specific requirement. If the autoclave temperature falls below the sterilizing temperature the melt solidifies and so an incomplete sterilization is recorded.

Of the devices which change colour during sterilization, Browne's tubes (A. Browne Ltd., Leicester) have gained wide acceptance. These small glass tubes contain a red fluid which changes colour through amber to green on heating. Three types are available which respond to (1) 115° for 25 min, (2) 121° for 15 min, (3) 160° for 1 h. It must be remembered that the colour change can occur with wet or dry heat and that because the reaction is temperature dependent some change can take place if the tubes are not stored cool. The safety margin with these tubes is satisfactory at $>110°$.

Chemically treated paper strips which change colour or darken are also available in this country. The purple Klintex paper (R. Whitelaw Ltd., Newcastle upon Tyne) loses its colour in the presence of steam so that the word AUTOCLAVED becomes visible in black against a white background. Unfortunately the end point is readily attained by conditions which will not kill all bacteria, and so the papers can only be used as evidence that a load has been exposed to heat.

A tape which develops an intensity of contrasts between light grey on a fawn background to almost black on the same background is also available commercially (3M Company London). Again, as with the Klintex strip, it cannot be used as an absolute measure of sterilization. It is useful, however, to assess the uniformity of heat treatment in a load. For this purpose the tape is stuck diagonally across each package and the uniformity of the colour change observed (Bowie, Kelsey and Thompson, 1963). Being an adhesive tape it can serve as a permanent indicator on the outside of a package to ensure that the package has been through a sterilization process.

Ethylene oxide sterilization indicators

Both chemical and biological indicators are available for use in ethylene oxide sterilization. As with steam sterilizations, chemical indicators have their limitations. They are insensitive to conditions of humidity and therefore do not necessarily indicate that successful sterilization has been attained: they simply show that the load has been exposed to the given concentration of ethylene oxide for the required time.

Several types of chemical indicators have been devised and chemical-treated paper tapes which change colour in the presence of the gas can be obtained commercially. For many years we have satisfactorily employed the Royce's sachet, developed in these laboratories (Royce and Bowler, 1959; Royce, 1960). In the sachet, of specific dimensions and polythene thickness, is sealed a mixture of saturated magnesium chloride solution and a controlled amount of hydrochloric acid with bromophenol blue indicator. During treatment, the ethylene oxide diffuses into the sachet, is absorbed by the solution to form ethylene chlorhydrin and the resulting alkaline reaction turns the indicator blue (Fig. 5). By varying the dimensions of the sachet and the quantity and strength of acid a number of sterilizing conditions can be controlled.

For the biological control of ethylene oxide sterilization Beeby and Whitehouse (1965) have described a method of producing spore strips which have found wide acceptance. They utilized spores of *Bacillus subtilis* var. *niger* (*B. globigii*) which are highly resistant to ethylene oxide and are stable on storage for several months when suspended in methyl alcohol. *Bacillus stearothermophilus* spores, have also been recommended, but these are rather less resistant to ethylene oxide.

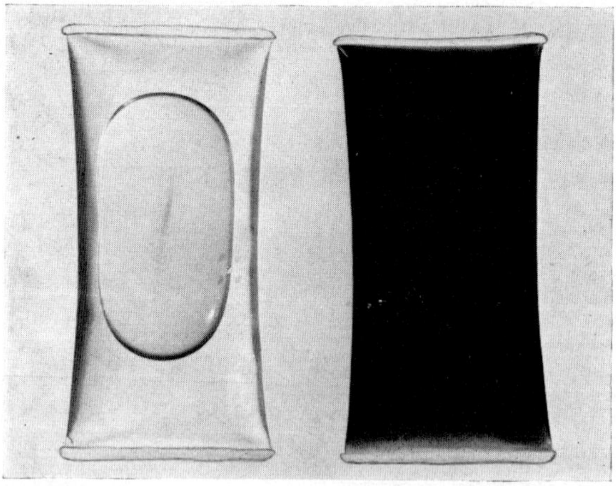

FIG. 5. The Royce's sachet. Left, before exposure to ethylene oxide (yellow); right, after a successful exposure to ethylene oxide (purple).

Tests for Sterility

Having been satisfied that the material has been properly sterilized and that the correct aseptic manipulative processes have been used, the final product can now be submitted to the last event in the chain of control, the test for sterility. This involves taking random samples from each batch of each day's work, the number depending on the size of the batch. The official tests in this country, as given in Therapeutic Substances Regulations and in the British Pharmacopoeia, at present specify 2% of the containers in the batch, or 20 containers, whichever is the fewer.

Each sample is examined for the presence of aerobic and anaerobic bacteria and in some instances also for moulds, although this is not specified in the official tests. The amount of each sample to be examined is 2 ml (1 ml aerobic and 1 ml anaerobic) of a liquid or 100 mg of a solid, but again it is not uncommon for much larger amounts to be examined.

The medium to be used can be either one which will detect aerobic and another which will detect anaerobic bacteria, or it can be a single medium capable of initiating and maintaining the growth of both aerobic and anaerobic bacteria. An ordinary Nutrient Broth and a Robertson's meat medium or a modified Brewer's medium with cysteine and thioglycollic acid are the most commonly employed, along with a modified Sabouraud's liquid medium or a Glucose Broth for the moulds test. Incubation is at 30–32° for 7 days, or at 25° for 10 days in the test for moulds.

It used to be the practice to inoculate the test sample into a volume of culture medium sufficient to dilute out any preservative or other antimicrobial activity in the preparation, and this was particularly difficult with antibiotics. The difficulty has now been overcome with the introduction of the membrane filter technique, although the manipulations are more hazardous microbiologically. In this technique the antibiotic, or any other drug preparation, is either filtered directly, or first dissolved in sterile water and then filtered, through a presterilized membrane filter. During filtration any organisms present in the sample are collected on the membrane and are so separated from the antibiotic or other drug, thus removing any inhibitory activity present in the preparation. The membrane, therefore, after being washed through with sterile water or saline, can be cultured directly into the test media. The technique is not applicable, of course, to insoluble drugs or to drugs in aqueous or oily suspension. It is the method of choice in our laboratories.

Because the operations of the test involve aseptic transfers to nutrient media there is always the hazard of adventitious contamination occurring, thus giving false positive results. A point of prime importance, then, is the elimination of such a hazard and so of establishing confidence in the test method, thus enabling action to be taken at the first indication of real contamination. Various sterile room techniques, with sterile clothing, sterile cabinets and directional airflow systems have been recommended, but our choice is the ethylene oxide-sealed screen method as devised by

FIG. 6. The ethylene oxide screen in use (front view).

FIG. 7. The ethylene oxide screen in use (view from the back port into the screen).

Royce and Sykes (1955). The method is reliable and virtually fool-proof, and can be operated with a considerable reduction in staff compared with the more traditional methods.

The screen (Figs 6 and 7) is made of tinned steel and is fitted with a Perspex window and various ports: when sealed it is absolutely airtight. Because of this and because the only access for work is *via* the gauntlet gloves, complicated manipulations can be carried out within the screen without any risk of accidental contamination being introduced.

In operation, the screen is loaded during the afternoon with all the necessary samples, media, syringes and other needed equipment, each item being carefully wiped over with 50% alcohol to ensure that the surfaces are clean. The last item to be introduced is a small, chilled bottle containing sufficient liquid ethylene oxide to give a gaseous concentration within the screen of *c.* 300 mg/1 (15% v/v). Having sealed the screen, the air in it is replaced, *via* the attached air filters, with nitrogen (to eliminate the explosive risk with ethylene oxide gas) and then the liquid ethylene oxide is released and allowed to evaporate. The sterilization is controlled by Royce's sachets which change colour after an exposure to a gas concentration of 250 mg/1 for 18 h at $\nless 20°$. The amount of ethylene oxide introduced is in excess of this to allow for loss by absorption by the rubber gauntlets.

At the end of the sterilization period the ethylene oxide in the screen is replaced by flushing through the air filters first with nitrogen and then with air. After this the gauntlets are left hanging outside the screen for about 1 h to allow all traces of gas to be desorbed from them, and then the screen is ready for use. Occasional swabs and blank control tests are needed to check the efficiency of the sterilization procedure.

Obviously all media must be in screwcapped bottles and all samples must be in ethylene oxide-proof containers. In general, there is no difficulty in sterilizing all of the surfaces within the screen, including those of the ancillary equipment, but syringes with tightly fitting plungers present an extra problem, and so it is preferable to use those which have been presterilized in steam. The membrane filtration method can be used with assurance in the screen, and the washing process, addition of an inactivator or the adjustment of a pH value present no problem.

References

BEEBY, M. M. & WHITEHOUSE, G. E. (1965). A bacterial spore test piece for the control of ethylene oxide sterilizations. *J, appl. Bact.*, **28,** 349.

BERNARD, H. R., SPEERS, R., O'GRADY, F. & SHOOTER, R. A. (1965). Reduction of dissemination of skin bacteria by modification of operating-room clothing and by ultraviolet irradiation. *Lancet*, **1,** 458.

BETHUNE, D. W., BLOWERS, R., PARKER, M. & PASK, E. A. (1965). Dispersal of *Staphylococcus aureus* by patients and surgical staff. *Lancet*, **1,** 480.

Blowers, R. & McCluskey, M. (1965). Design of operating-room dress for surgeons. *Lancet*, **2,** 681.

Bowie, J. H., Kelsey, J. C. & Thompson, G. R. (1963). The Bowie and Dick autoclave tape test. *Lancet*, **1,** 586.

Brewer, J. H. & McLaughlin, C. B. (1954). A device for determining time and temperature of sterilization in the autoclave or hot-air oven. *Science, N.Y.*, **120,** 501.

Kelsey, J. C. (1958). The testing of sterilizers. *Lancet*, **1,** 306.

Kelsey, J. C. (1961). The testing of sterilizers. 2. Thermophilic spore papers. *J. clin. Path.*, **14,** 313.

Perkins, J. J. (1969). *Principles and Methods of Sterilization in Health Sciences.* Springfield, Illinois: Thomas.

Royce, A. (1960). Brit. Pat. 854142.

Royce, A. & Bowler, C. (1959). An indicator control device for ethylene oxide sterilization. *J. Pharm. Pharmac.*, **11,** 294T.

Royce, A. & Sykes, G. (1955). A new approach to sterility testing. *J. Pharm. Pharmac.*, **7,** 1046.

Spooner, E. T. C. & Turnbull, L. H. (1942). Notes on the use of autoclaves; with special reference to the sterilization of blood transfusion apparatus. *Bull. War Med.*, **2,** 345.

Sykes, G. (1965). *Disinfection and Sterilization*, 2nd ed. London: Spon.

Sykes, G. (1968). *Control and sampling in sterile rooms.* COSPAR Tech. Man. Ser., Manual No. 4 (P.H.A. Sneath, ed). Paris: COSPAR.

The Isolation and Identification of Bacteria and Mycoplasma Pathogenic to Laboratory Animals

D. K. Blackmore and A. C. Hill

M.R.C. Laboratory Animals Centre, Carshalton, Surrey, England

Any competent pathology laboratory should find no major difficulty in isolating and identifying organisms associated with frank disease. Obviously, a diagnostic laboratory would be aware of those pathogens usually associated with the species of animal under examination, and in most cases of such diagnostic microbiology, the pathogen is present in relatively large numbers, and in certain circumstances may be the only organism isolated. The heart blood of a pig which has died of acute erysipelas is likely to reveal a pure growth of *Erysipelothrix rhusiopathiae*. An animal which has died as a direct result of enteric Salmonellosis is likely to have large numbers of the organism in the alimentary tract, which, in spite of the presence

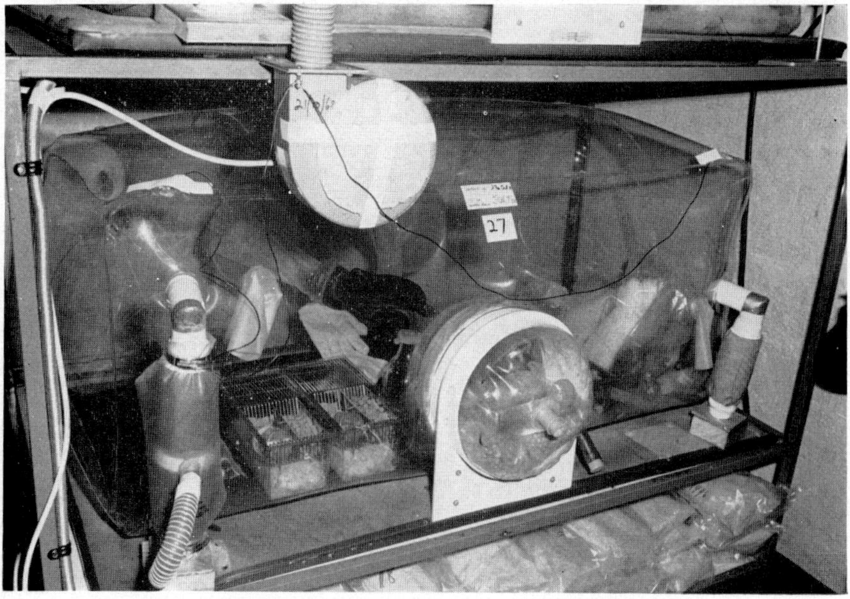

Fig. 1. An isolator made from plastic film.

of non-pathogenic enteric organisms, will be easily demonstrated with the aid of selective media. If examination *post-mortem* of a rabbit reveals an acute pneumonia, from which large numbers of a pasteurella-like organism are isolated, the cause of death and aetiology may sometimes be assumed without exact and detailed classification of the causal organism. The microbiological screening of clinically healthy animals often requires a very

FIG. 2. Servicing an isolator.

different approach to the methods of isolation and the extent of determinative bacteriology.

The increasing use of barrier-maintained—typical isolators shown in Figs 1 and 2—and specific pathogen-free (SPF) animals has inevitably resulted in more laboratories becoming involved with the routine screening of such colonies. The random screening of animals for potential pathogens, or other organisms which would indicate a break-down of the barrier system, plays an integral part in the maintenance of such colonies. As the animals selected for examination are likely to be clinically healthy, on culture the normal non-pathogenic flora will predominate, making the isolation of pathogens more difficult, and necessitating rather more meticulous isolation procedures than those required for a diagnostic autopsy.

This contribution is designed as an aid to the microbiologist who becomes involved with the screening of laboratory animals without having much previous experience in this sphere of work. It is not designed as a treatise on determinative bacteriology, and assumes that the reader is conversant with such techniques and standard text books on the subject. The information and suggestions are based on the experience gained by the authors at the M.R.C. Laboratory Animals Centre over the course of several years, during which time many thousands of animals have been examined. It is hoped that these suggestions will assist others in devising techniques suitable for screening their own colonies of animals and adapted to provide the information required for their own purposes. Although this chapter is essentially concerned with bacteria and mycoplasma, infections due to viruses and higher parasites are also of paramount importance. It is hoped that a much more comprehensive handbook on the parasites of laboratory animals will be published separately (Owen, in press). This Centre has only recently begun to investigate the problems of screening animals for evidence of virus infections, and when more progress has been made, suggestions will be published.

General Considerations

Before instigating an animal screening system, there are five major considerations to be taken into account and appropriate decisions to be reached.

1. Specification of those organisms considered to be potentially pathogenic to the species of animal being examined, and those organisms whose presence would indicate faulty barrier management.

2. Examination of the sites most suitable to demonstrate these specified organisms within the animal body.

3. The most practical microbiological procedures to isolate and identify suspect organisms.

4. The choice of an adequate but practical sample size, and the frequency of such sampling.

5. Elimination of errors associated with either contamination in transit or in the laboratory, or confusion between different samples submitted for examination.

Specified organisms

Those organisms which will be considered undesirable will vary according to the host and the purpose for which the animals are required. In all circumstances, salmonellae must presumably be considered a specified pathogen, while *Pseudomonas aeruginosa* would not necessarily be a specified pathogen in mice, unless the animals were required for work involving whole body irradiation.

Table 1 gives a list of organisms that are considered to be common pathogens or potential pathogens for the more common laboratory rodents and lagomorphs, which should be able to be demonstrated relatively easily by the majority of bacteriology laboratories. This list is not considered absolute, and some obviously pathogenic organisms, such as leptospira, have been omitted. As mentioned previously, any such list of specified pathogens will depend on the purpose for which the animals are required and techniques which are within the capability of the laboratory screening the animals.

TABLE 1. Bacteria and mycoplasmas specified as pathogens and demonstrable by direct methods

Bordetella bronchiseptica	*Mycoplasma pulmonis*
Corynebacterium kutscheri (*muris*)	*Pastuerella multocida*
Diplococcus pneumoniae	*Pasteurella pneumotropica*
Erysipelothrix rhusiopathiae	*Salmonella* (all species)
Klebsiella pneumoniae	*Shigella* (all species)
Listeria monocytogenes	*Streptobacillus moniliformis*
Mycobacterium tuberculosis	*Treponema cuniculi*
Mycoplasma arthritidis	*Yersinia enterocolitica*
Mycoplasma neurolyticum	*Yersinia pseudotuberculosis*

Sites of isolation from animal

The routine examination of animals from a barrier-maintained building is a form of quality-control testing. It is therefore, important that the methods should be as constant as possible, and the tests reproduceable, so that a meaningful and realistic standard may be achieved.

These sites of culture are of two main types; those which are always examined, irrespective of the general *post-mortem* findings, and those which are examined because a definite lesion is present. For instance, the nasopharynx of a rat will be found to be harbouring *Mycoplasma pulmonis*, whether the rat is a non-clinical carrier or is suffering from chronic respiratory disease. On the other hand, if an animal is affected with tuberculosis, *Mycobacterium tuberculosis* is only likely to be demonstrated from the actual tubercles or other lesions. It should be remembered that quite a number of the bacterial pathogens of laboratory animals can, on occasion, produce disease in man, and precautions must be taken to avoid infection from pathological specimens or cultures. In particular, cultures of *Mycobacterium tuberculosis* (human) and *Shigella* and *Salmonella* spp should be treated with caution. A complete section will be devoted to the methods of primary isolation of organisms from a carcase.

Identification of organisms isolated

The identification and classification of any organism is dependent on acknowledged principles which are outlined in appropriate reference books (Wilson and Miles, 1964; Cowan and Steel, 1966). This chapter does not intend to supersede these standard references, but merely to outline some of the more important criteria which have been found to be particularly useful in helping to classify the pathogens of laboratory animals.

Various strains of pathogenic organisms may not exactly conform to the biochemical criteria normally used for identification. It is unlikely that any one laboratory is experienced in the culture and identification of all the organisms listed in Table 1. It will, therefore, sometimes be necessary to seek the opinion of independent experts or reference laboratories, before reaching a decision which might profoundly affect the status of a colony. A complete section will be devoted to these problems of identification and classification.

Sample size and frequency of testing

It is not possible to lay down rigid recommendations for either the sample size of animals for screening, or for the frequency of such tests. Factors governing such decisions include the type and size of the animal colony and the characteristics of the organisms in which one is particularly interested. The period between tests is a particularly arbitrary decision, dependent on the purposes for which the animals are to be used. Some animal colonies are only tested every six months, while weekly samples are examined from the Centre's own SPF unit. In many cases, monthly samples might be a realistic compromise. If, however, a barrier-maintained unit has been subjected to a particular hazard, such as a power failure or autoclave breakdown, it may be considered necessary, for a time, to increase the frequency of animal testing.

The actual sample size is primarily dependent on the availability and economic value of animals within the unit. It would be impracticable to cull large numbers of cats each month, but the value and availability of mice should seldom create a serious problem. If a barrier building is sub-divided into self-contained units, each sub-division may have to be considered as a separate entity, while if the unit is only sub-divided into inter-communicating animal rooms, the whole building could be treated as one unit for the purposes of sampling.

Obviously, the larger the sample size, the more accurate the results obtained. Certain organisms, such as the pasteurellae and the higher parasites, with a direct method of transmission, spread very rapidly through an animal colony, affecting the majority of the population, and there is a high probability of detecting such infections on relatively small samples. Conversely, other organisms, such as *Listeria monocytogenes*, probably do not spread so rapidly, and without a sample size which would be economically unrealistic, definite detection of such an infection would be unlikely on an individual sample. However, as the life of a barrier-maintained colony should be several years, the cumulative information, gained from an increasing total of samples examined, gives a more accurate assessment.

At the Laboratory Animals Centre, the usual sample is eight animals. Four adults are submitted for bacteriological examination, and two adults and two young animals are submitted for a detailed parasitological examination.

Elimination of error

In any system, errors can occur; in a well run laboratory the chances of such an occurrence should be minimal. With reference to the screening of laboratory animals, there are two main areas of hazard: (1) during the transit of samples between the animal house and laboratory, and (2) in the laboratory, by contamination of culture sites or media, or by mislabelling of samples. If animals are transported in pre-sterilized filter-top boxes, the chance of contamination during transit should be remote (see Fig. 3). The elimination of contamination during actual laboratory procedures depends on proper laboratory discipline, and on a fully trained technical staff conversant with aseptic techniques. Great care must be taken in the identification of separate samples, and the use of a sound reference system. Final close scrutiny of all results by the head of department will also help to avoid the presentation of results which are unexpected, without first rechecking on all the techniques employed during the examination.

FIG. 3. Filter-top boxes.

Autopsy and Isolation Techniques

Before the animal is destroyed, its behaviour should be observed, and any signs of infection noted. Apart from pre-decided culture sites, any other organ or lesion seen to be infected during a thorough autopsy should be included in any screening routine.

Euthanasia and preliminary examinations

Destruction in a carbon dioxide chamber appears to result in a minimum of terminal changes in small rodents; rabbits and larger animals are destroyed by intravenous sodium pentobarbitone. The animal should be examined within 30 min of death. In all species, the exterior of the animal is carefully examined for signs of cutaneous diseases. If a lesion is indicative of a dermatrophic fungal infection, hair and skin from the affected area are treated with 10% (w/v) potassium hydroxide solution and examined microscopically. Affected tissue is then taken for histological examination, and cultures are made on Sabouraud's agar which is incubated at room temperature. If cultures are positive, they are sent to a specialist laboratory for identification.

The animal is secured in dorsal recumbency, and its ventral surface swabbed with a germicidal agent (2% Hycolin) to lessen the risk of hair or other external debris contaminating subsequent culture sites. The skin from the ventral surface of the abdomen, thorax and neck is reflected, and the abdomen opened with sterile instruments. A small, electrically heated water sterilizer, in which instruments are kept between each operation, has been found most useful for this purpose.

Cultures

Culture of gut and liver

The caecum is exposed, and, for the isolation of salmonellae and shigellae, a loopful of content is plated on to a Desoxycholate Citrate Agar (DCA) plate, and a further loopful transferred to a Selenite F Broth. If a gall bladder is present, a loopful of bile is transferred to a separate DCA plate, and the same Selenite Broth. If no gall bladder is present, direct liver cultures are used *in lieu*. After 24 h incubation at 37°, subcultures are made from Selenite Broth to DCA plates. The DCA plates are incubated for 48 h at 37°.

To demonstrate possible generalized bacterial diseases, cultures of liver are taken aseptically, and transferred to a Robertson's cooked meat medium, a 10% (v/v) Horse Blood Agar plate, and a MacConkey Agar plate. The cooked meat medium is incubated for 48 h at 37°, before being sub-cultured on to Blood and MacConkey Agar plates. It is also useful to examine a Gram-stained smear from the cooked meat medium, and should this or other observations suggest that anaerobes are present, the cultures should be transferred to a further Blood Agar plate incubated anaerobically. These primary and secondary plates are examined daily for 2 days before being discarded. The remainder of the abdominal organs are now examined, before proceeding with culture and examination of the thoracic organs.

Culture of respiratory tract

In all species, cultures of the hilus of the lung are made directly on to 10% (v/v) Horse Blood Agar, MacConkey Agar and Robertson's cooked meat media. The agar plates are incubated at 37° for 48 h. The cooked meat medium is incubated for 48 h before being sub-cultured on to Blood and MacConkey plates, which are again incubated for a further 48 h. Such procedures are expected to demonstrate pneumotrophic bacteria. Myocplasma cultures are not made from the lungs unless obvious lesions are present, as experience has shown that such organisms are also inevitably present in the naso-pharynx of rats or mice, if such a respiratory infection associated with mycoplasma exists.

In all species, the naso-pharynx is also cultured on to Blood and MacConkey plates and 30% (v/v) Horse Serum Agar. The Blood and MacConkey plates are incubated for 48 h and the 30% (v/v) Serum Agar is incubated for 4 days. This latter medium is used to demonstrate *Streptobacillus moniliformis*. Serum Broth can also be a most useful enrichment medium for many aerobes. The cultures are taken from the pharynx by removing the larynx, and then passing a cottonwool swab dorsal to the soft palate into the naso-pharynx, and forward into the turbinate region, where it is revolved and then removed. In the case of rats and mice, cultures are also made on to appropriate mycoplasma plates and broths, which are incubated for three weeks before being discarded. Medium used for isolating mycoplasma is that described by Taylor-Robinson, Williams and Haig (1968). Material from rats and mice is inoculated on to plates and into 2 liquid media, containing 0·1% (w/v) glucose and 0·1% (w/v) arginine respectively. The glucose medium will be fermented by *Mycoplasma pulmonis* and *M. neurolyticum*, while the arginine will be utilized by *M. arthritidis*.

Culture of middle ear

The animal is unpinned from the autopsy board and the skin overlying the skull reflected, so that the pinna is removed as close to the tympanic membrane as possible. The tympanic membrane is ruptured with a sterile blunt metal probe, and by means of a Pasteur pipette, a few drops of sterile saline flushed several times in and out of the middle ear. This fluid is inoculated on to 30% (v/v) Serum Agar and, in the case of rats and mice, appropriate mycoplasma plates and broth. These media are incubated for 4 days and three weeks respectively.

Identification of Organisms

Before outlining the more important criteria which are used to identify pathogens isolated, it must be reiterated that this information is only designed as a guide for those who are already conversant with the culture and identification of microorganisms, and who have some basic knowledge of the organisms described. Some experience is required before mycoplasmas can be cultured efficiently, and screening of animals for these and other fastidious organisms should not be attempted until such experience has been attained.

It must also be remembered that organisms apparently associated with certain infectious diseases may be extremely difficult or impossible to isolate. For instance, Tyzzer's disease of mice can only be confirmed by histological examination. On the other hand, although *Fusiformis necrophorus* is associated with labial ulceration of the rabbit, it is questionable whether exposure of a rabbit colony to this organism alone will result in an epidemic of the disease and, indeed, it has been stated that this organism is a normal inhabitant of the rabbit's skin (Ostler, 1961). It would seem, therefore, that *F. necrophorus* should only be considered a significant pathogen if associated with a definite lesion.

In the subsequent lists of presumptive criteria, the usual reactions on various carbohydrate media are given. It is considered that the following media should form the basis for evaluating such tests: glucose, lactose, sucrose, mannite, salicin, dulcite, maltose, glycerol, sorbitol, xylose, and aesculin. Each carbohydrate being in the form of a 1% solution in peptone water at a pH of 7·6–7·8 and with added Andrades indicator. However, in many cases information on all these fermentative reactions will not be required. Obvious basic data such as colonial characteristics and Gram staining reactions are not included, as it is assumed that anyone considering this type of work would be conversant with such elementary details.

Characters used for identification

Bordetella bronchiseptica

B. bronchiseptica is motile and urea positive, but produces no acid or gas in any of the carbohydrate media mentioned above. Final presumptive evidence should be obtained by guinea-pig inoculation. One ml of an 18 h broth culture is injected by the intraperitoneal route, and the animal should die within 24–48 h. At *post mortem* examination, the animals should exhibit a typical septicaemic appearance with a viscid translucent peritoneal exudate from which the organism can be recovered.

Corynebacterium kutscheri (muris)

C. kutscheri is non-motile, catalase positive and reduces nitrates. Acid is produced in glucose, maltose and sucrose, but not from lactose or mannite, and there is no gas production in any carbohydrate media. *C. kutscheri* appears to be a relatively uncommon organism, and final diagnosis may depend on the opinion of a reference laboratory.

Erysipelothrix rhusiopathiae

E. rhusiopathiae is non-motile, catalase negative, and may reduce nitrates weakly. Acid, but no gas is produced in glucose and lactose, but not from sucrose. If serological facilities are not available, mice should be inoculated intraperitoneally with 0·5 ml of an 18 h broth culture, which should prove fatal within 2 to 3 days.

Klebsiella pneumoniae

K. pneumoniae is non-motile and should produce typical mucoid colonies on both Blood and MacConkey Agar. It is a lactose fermenter, can utilize citrate, is indole negative, methyl red positive, and Voges Proskauer negative. On intraperitoneal injection of mice with 0·5 ml 18 h broth culture, death should occur within 24 h.

Listeria monocytogenes

L. monocytogenes will probably rarely be encountered during the routine screening of laboratory animals. The organism grows on both Blood and MacConkey Agar, is catalase positive, exhibits a tumbling motility at 20–25°, is capable of growth at 4°, and fails to reduce nitrates. Colonies on blood agar show a narrow zone of haemolysis. Instillation of a culture into the conjunctiva of rabbits should result in a severe conjunctivitis within 24 h, followed by a keratitis.

Mycobacterium tuberculosis (*human and bovine strains*)

M. tuberculosis from lesions from which typical acid-fast organisms have been demonstrated should be cultured directly, and after treatment by Petroff's method, on to Dorset egg medium. Material similar to that inoculated on to egg medium should also be inoculated sub-cutaneously into guinea-pigs, and the subsequent development of tuberculosis by the animals confirmed.

Mycoplasma

M. pulmonis and *M. neurolyticum* produce acid from glucose and fail to utilize arginine, while *M. arthritidis* utilizes arginine and fails to produce acid from glucose. Final identification must depend on results of disc inhibition techniques, metabolic inhibition or complement fixation tests.

Pasteurella multocida

P. multocida is non-motile, indole positive and urea negative. Acid, but no gas, is produced in mannite but not from lactose and glycerol. Diagnosis is confirmed by the intraperitoneal inoculation of mice with 0·5 ml of an 18 h broth culture, which should cause death within 3 days.

Pasteurella pneumotropica

P. pneumotropica is non-motile and urea positive. Acid but no gas is produced by fermentation in glucose, maltose, sucrose and glycerol, but not from mannite or sorbitol. Animal inoculation does not aid diagnosis, as it may be difficult by normal experimental methods, to demonstrate obvious pathogenicity.

Diplococcus pneumoniae (*pneumococci*)

Pneumococci usually exhibit a characteristic appearance on blood agar, are catalase negative, bile soluble and sensitive to optochin. Intraperitoneal inoculation of mice with 0·25 ml of an 18 h serum broth culture should cause death within 24 h.

Salmonella *spp*

Salmonellae are motile organisms, which produce acid and gas from glucose and mannite, but not from sucrose or lactose. There are a few notable exceptions to these criteria (*S. pullorum* is non-motile; *S. typhi* is anaerogenic), but these are unlikely to occur in laboratory rodents. Salmonellae are indole and urea negative. Slide agglutination tests sould be carried out using at least polyvalent salmonella O and H antigens. Final confirmatory diagnosis will usually depend on submission of cultures to the Salmonella Reference Laboratory.

Shigella *spp*

Shigellae are unlikely to be recovered, unless primates are being examined. They are non-motile organisms, urea negative, and rarely ferment carbohydrate media with gas production. All strains produce acid from glucose, some ferment mannite and some are late fermenters of lactose and sucrose. Suspect organisms should be subjected to slide agglutination tests, using group specific shigella antisera. Final confirmation should depend on the submission of cultures to a reference laboratory.

Streptobacillus moniliformis

S. moniliformis is a relatively fastidious organism, which grows best under anaerobic conditions, preferably with added CO_2 (10%), though some growth may occur aerobically; growth is also favoured by a humid atmosphere. It is unable to grow on plain agar, but should produce small colonies on 30% (v/v) Serum Agar within 48 h. The organism is pleomorphic with filamentous forms, which often show fusiform, oval, spherical or club-shaped swellings. Presumptive diagnosis must depend on the sub-cutaneous inoculation of the foot pads of mice with a serum broth culture of the organism. Marked swelling of the feet, associated with an arthritis from which the organism can be recovered, should occur within three weeks.

Treponema cuniculi

Presumptive evidence for the presence of *Treponema cuniculi* is based on the demonstration, by dark ground microscopy, of typical spirochaetes from lesions of the genital tract of rabbits, which are typical of those associated with rabbit syphilis. Cultural and serological tests are usually beyond the scope of a laboratory which does not specialize in this type of organism.

Yersinia pseudotuberculosis

Y. pseudotuberculosis is motile at 22°, urea positive and indole negative. Acid, but not gas, is produced in mannite, salicin, glycerol, xylose and aesculin, and either late or not at all from sucrose. No acid is produced from lactose. A final diagnosis is obtained by subcutaneous inoculation of guinea-pigs with 1 ml of an 18 h broth culture. Animals which do not die within 3 weeks from an acute form of the disease should exhibit typical lesions of pseudotuberculosis at examination *post mortem*.

Yersinia enterocolitica

Y. enterocolitica is somewhat similar to *Y. pseudotuberculosis*, being motile at 22°, urea positive and indole negative. It does not produce gas from any carbohydrate medium, nor acid from lactose. However, it will not produce acid in xylose or aesculin.

Microbiological Control of Germfree Animals

A germfree animal is only as "germfree" as the sophistication of the tests to which it is subjected. Ideally, animals should be regularly screened by as many tests as possible, which are likely to detect evidence of contamination by any organism including viruses, but such extensive tests are usually beyond the scope of most laboratories. The more usual approach is to check fresh faeces and other isolator waste on a regular basis for evidence of contamination by bacteria or higher parasites, and to subject animals to much more extensive tests when the opportunity arises, so that over a period of time a more comprehensive "microbiological pedigree" of the colony is established. Apart from information gained directly from microbiological examinations, most useful indirect evidence of the animals' gnotobiotic state can be gained from macroscopic and histological examination of carcases. Such examination should confirm certain anatomical features which are associated with the germfree state, such as an enlarged caecum (Figs 4 and 5), a characteristic gut mucosa and an underdeveloped reticulo-endothelial system.

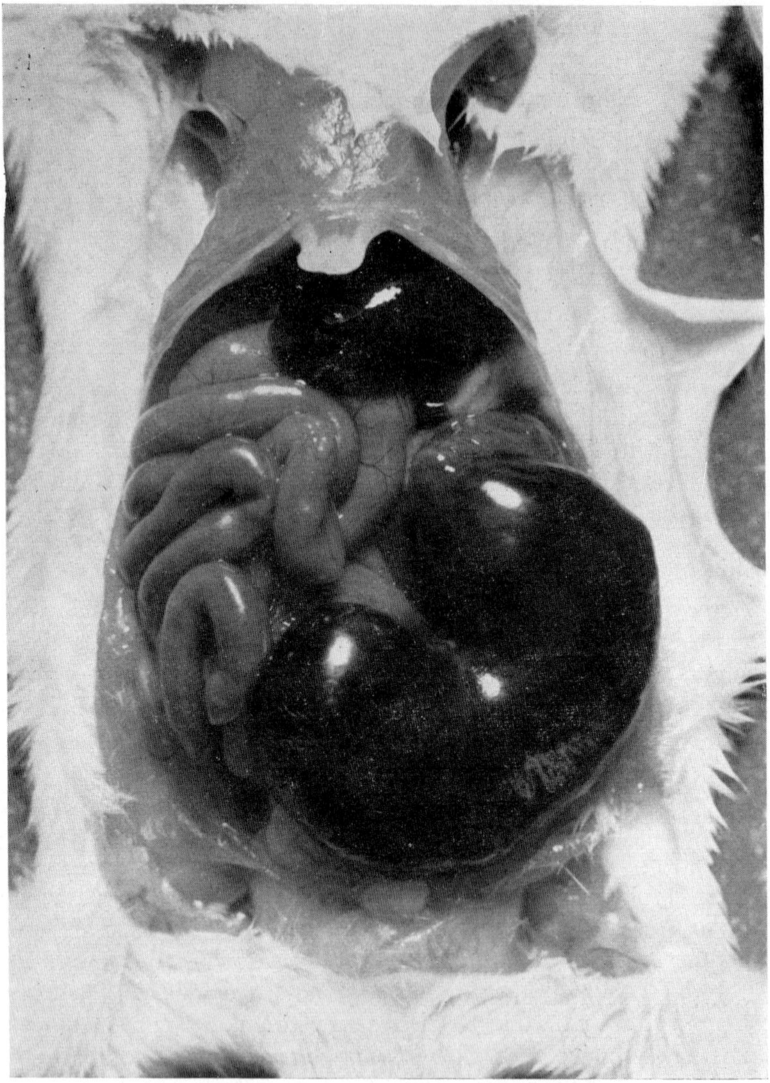

FIG. 4. Large caecum of germ-free animal.

At the Laboratory Animals Centre, faeces from animals in each isolator are screened regularly each week. Other types of screening procedures are carried out only when excess animals or specialized techniques are available.

Faeces are collected directly in a sterile bottle from the rectum (Fig. 6) of

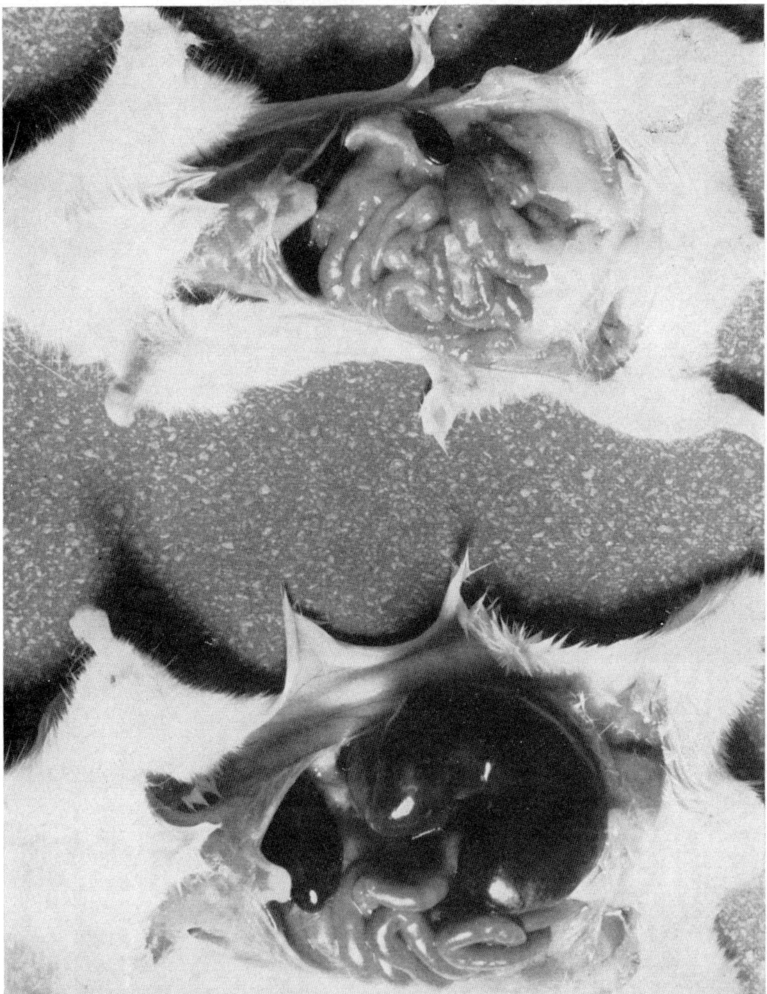

FIG. 5. The viscera of a germ-free (left) and a conventional (right) animal.

FIG. 6. Taking a faecal sample from a rat held in an isolator.

a few animals in each isolator, and immediately despatched to the laboratory where all further tests are carried out in a laminar horizontal flow cabinet—for details, see p. 21. The following tests are carried out:

Direct examination

Preparations are emulsified in normal saline and examined for the presence of motile bacteria and higher parasites. Direct smears are stained with a vital dye (Watson's stain, see Stitt *et al.*, 1948), and examined for evidence of viable organisms. Non-viable organisms originating from the sterilized diet are usually present.

Culture techniques

Faeces are cultured in a variety of media, at room temperature, 37° and 56°, and under aerobic, anaerobic and microaerophilic conditions.

Media inoculated

The following media are initially inoculated:

Robertson's cooked meat medium	1*
Thioglycollate Broth	3
Brain-Heart Infusion Broth	4
Sabouraud's liquid medium	1
Heated Blood Agar plates	3

Temperatures of incubation

The following media are incubated for three weeks at the temperatures indicated:

25°	Thioglycollate Broth	1*
	Brain-Heart Broth	1
	Sabouraud's medium	1
37°	Thioglycollate Broth	1
	Cooked meat medium	1
	Brain-Heart Broth	2
56°	Thioglycollate Broth	1
	Brain-Heart Infusion	1

Sub-cultures and heated blood agar plates

The cultures incubated at 25°, 56°, the Thioglycollate Broth and one of the Brain-Heart Infusion Broths incubated at 37° are only examined for evidence of turbidity.

* No. inoculated.

The three primarily inoculated Heated Blood Agar plates are incubated at 37° under aerobic, anaerobic and microaerophilic conditions, and kept as long as possible.

Subcultures are taken from the cooked meat medium and one of the Brain-Heart Infusion Broths, on to heated Blood Agar plates, and incubated respectively under anaerobic and aerobic conditions. These subcultures are taken at 48 h, 7 and 14 days. The plates are incubated for 4–5 days.

Periodically, complete animals are also examined bacteriologically, when media suitable for isolating mycoplasma are also incorporated, especially for cultures of the upper respiratory tract.

Acknowledgements

The authors wish to thank Professor J. E. Smith, BSc, PhD, MRCVS, for scrutinizing this Chapter and for his helpful comments.

References

COWAN, S. T. & STEEL, K. J. (1966). *Manual for the Identification of Medical Bacteria*. Cambridge: Cambridge University Press.

OSTLER, D. C. (1961). The diseases of Broiler Rabbits. *Vet. Rec.*, **73,** 1237.

OWEN, D. (in press). *Common Parasites of Laboratory Animals.*

STITT, E. R., CLOUGH, P. W. & BRANHAM, S. E. (1948). *Practical Bacteriology, Haematology & Parasitology*. 10th ed. London: H. K. Lewis.

TAYLOR-ROBINSON, D., WILLIAMS, M. H. & HAIG, D. D. (1968). The isolation and comparative biological and physical characteristics of T-mycoplasmas of cattle. *J. gen. Microbiol.*, **54,** 33.

WILSON, G. S. & MILES, A. A. (1964). *Topley and Wilson's Principles of Bacteriology and Immunity*. London: Edward Arnold.

The Production of Disease Free Embryos and Chicks

E. G. HARRY

Houghton Poultry Research Station, Houghton, Huntingdon England

AND

MARY MCCLINTOCK

Ministry of Agriculture, Fisheries and Food, Agricultural Development and Advisory Service, Coley Park, Reading, England

The production of healthy chicks has always been the aim of hatcheries providing replacement stock for poultry farms. Over the past few years however there has been an increasing demand for chick embryos and chicks which can be certified as being free from specified disease agents. These are required for the preparations of vaccines (Biggs, 1970*a*) and for various experimental purposes (Miner, 1952; New, 1966; Payne, 1967).

With embryos and chicks used for experimental purposes, freedom from disease, or at least some knowledge of the disease agents, or parenterally acquired antibodies likely to be present, is necessary if standard responses are to be obtained from their use. Where embryos are used for vaccine preparation contamination with extraneous bacteria or viruses is undesirable and in the case of live vaccines can have serious consequences (Zarger and Pomeroy, 1950; Hungerford, 1968).

In the case of parent birds infected with one of the egg transmissable disease agents, infection of embryos and chicks can result from infection of the eggs before they are laid (congenital infection). It can also result from microbial penetration of the shell following direct or indirect contact of the egg with faeces or other infected material in the nest box. Cross infection of chicks can also occur within the hatchery. The nature and significance of these sources of infection are discussed in the following pages, and methods are indicated whereby they can be eliminated or their effect minimized.

Congenital Infection

This has been shown to occur in the viral diseases: avian encephalomyelitis (AE) (Taylor, Lowry and Raggi, 1955), lymphoid leucosis (LL) (Burmester,

1962; Rubin, Cornelius and Fanshier, 1961), and infectious bronchitis (IB) (Fabricant and Levine, 1951). Egg and embryo infection with Newcastle disease (ND) has also been demonstrated (Van Roekel, 1946) but transmission to the chick is considered unlikely. Although it it possible that Marek's disease may also be transmitted through the egg there is strong evidence that this is unlikely (Biggs, 1970*b*). In addition to the disease associated viruses, egg transmission can also occur in the case of the viruses CELO and GAL and other adenoviruses which, although of uncertain pathogenicity in fowls, may have undesirable effects on the embryo and chick (Payne, 1968).

Certain bacterial diseases such as pullorum disease (Rettger and Plastridge, 1932), fowl typhoid (Jordan, 1956), salmonellosis, (Buxton, 1957) and mycoplasmosis (Fabricant *et al.*, 1959) can also be transmitted through the egg. *Mycobacterium tuberculosis* has also been isolated from eggs laid by tuberculous hens but this is considered to be rare (Fitch, Lubbehusen and Dikmans, 1924).

The nature of these diseases in the fowl is described by Biester and Schwarte (1965). The frequency with which eggs are likely to be infected varies with the disease concerned as indicated in the review of this subject by Payne (1968).

The Detection of Infection in the Laying Flock

In the case of vaccine production, eggs have to be taken from specified-pathogen-free (SPF) flocks (Chute, Stauffer and O'Meara, 1964). These are the concern of a limited number of licensed vaccine manufacturers and require specialized technical facilities—for additional details, see p. 103. These have been described by Biggs (1970*a*), Egan and Butler (1972*a*, *b*) and Cooper (1970) and will not be discussed further here. The disease status of flocks involved in providing embryos and chicks for experimental and general purposes, however, can be determined by regular tests on a proportion of the flock. In most cases these tests are based on the detection of serum antibodies to the various egg transmissable disease agents. In the case of the actual birds tested a positive correlation between the presence of antibodies and active infection leading to possible egg transmission only occurs in certain diseases (Payne, 1968). In most cases therefore the tests are used as a means of determining whether actively infected birds are likely to be present in the flock as a whole rather than a means of eliminating individual infected birds from the flock. The results will indicate whether or not it is advisable to take eggs from these flocks for incubation purposes.

Monitoring tests for detecting disease in the flock vary from the relatively simple stained bacterial antigen/serum plate tests, to the more complex tests required for the detection of viral infections. These include the haemagglutination inhibition test for Newcastle disease (Fig. 1), the embryo sensitivity test for avian encephalomyelitis (Fig. 2) and the serum neutralization

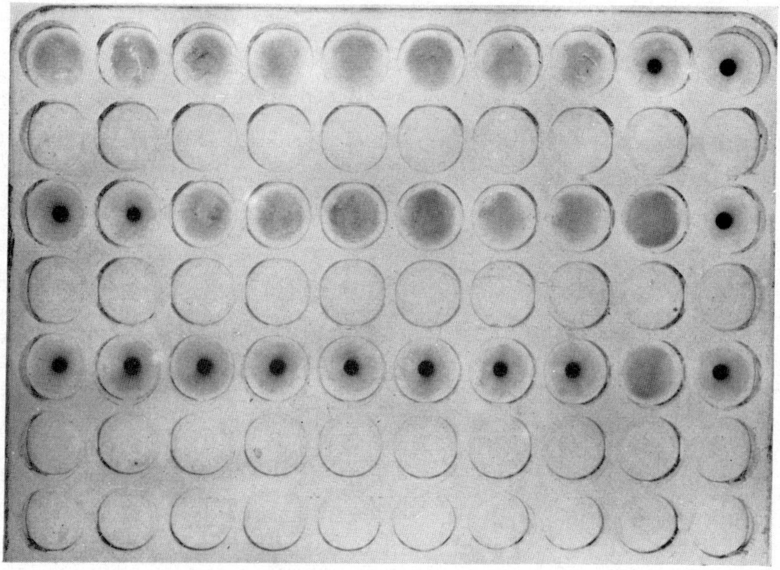

FIG. 1. Haemagglutination inhibition (HI) test for the presence of serum antibodies to Newcastle disease. Row 1. antigen titration—complete haemagglutination up to dilution 8; row 2. test serum dilutions + antigens—haemagglutination inhibited up to dilution 2, and row 3. positive control serum + antigen—haemagglutination inhibited up to dilution 8. Last wells in each row are red cell suspension controls showing absence of autoagglutination.

test for infectious bronchitis and lymphoid leucosis. Details of these tests are given by Grumbles *et al.* (1963). For the detection of Marek's disease a gel diffusion test (Chubb and Churchill, 1968) can be used.

Pullorum infection and fowl typhoid can be detected by means of a serum tube agglutination method or (Fig. 3) by the rapid stained antigen test (Gordon, 1942). *Mycoplasma gallisepticum* infection can also be detected by a stained antigen, a tube agglutination or an haemagglutination inhibition test (Leach and Blaxland, 1966).

Of the disease agents involved in shell contamination, *Salmonella typhimurium* infection can be detected in individual birds by serological tests coupled with rectal swabbing on 3 successive days (Gordon and Tucker, 1956). In the case of pens of birds it is often more convenient to culture the litter for the excreted salmonellae.

Birds found by these tests to be free from Pullorum infection, fowl

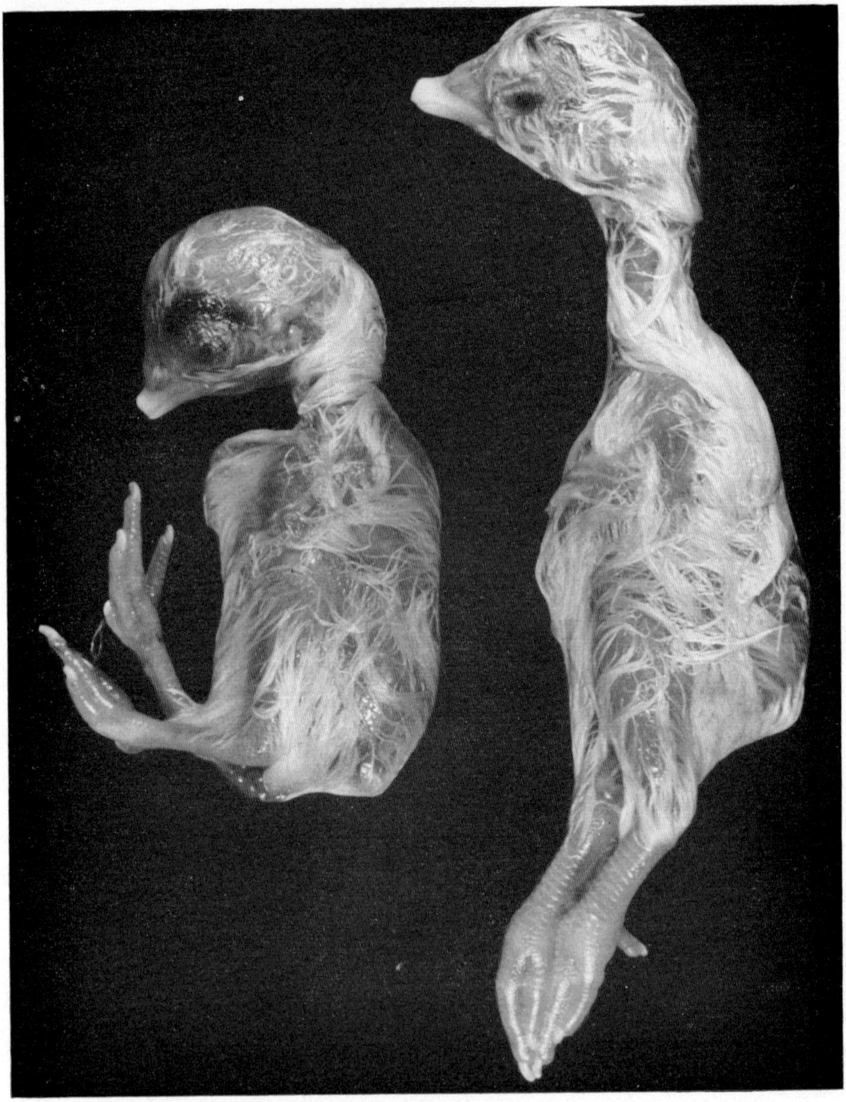

FIG. 2. Embryo sensitivity test for avian encephalomyelitis (AE) antibodies. Dwarfed embryo on left inoculated 10 days earlier with AE virus + negative serum. Normal embryo on right inoculated similarly with AE virus + positive serum.

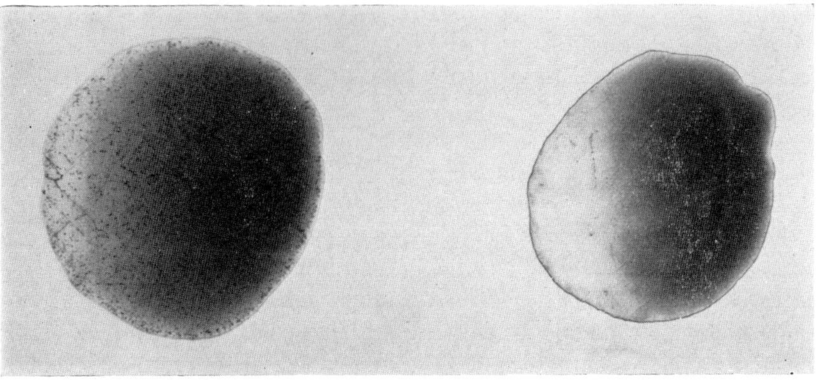

FIG. 3. *Salmonella pullorum* stained antigen plate test. Reaction after 1–2 min after mixing antigen and serum. Positive serum (clumping of antigen) shown on left.

typhoid or salmonellosis can be transferred to clean accommodation and retested later. By these means it is possible to eradicate these diseases from the flock.

Shell contamination and its control

Infection of the egg after it is laid can result from contamination of the shell, and subsequent penetration, by bacteria and fungi derived from faeces or decaying vegetable matter. These may include pathogenic bacteria such as *Salm. typhimurium* and *Salm. thompson* (Buxton and Gordon, 1947) and certain other salmonellae, certain types of *Escherichia* (Harry and Hemsley, 1965) and *Staphylococcus aureus* (Harry, 1957). Certain microbes, mainly saprophytic, such as *Bacillus cereus, Proteus* spp, pseudodomonads and *Aspergillus fumigatus* can also be lethal to the embryo or chick if present in sufficient numbers or if conditions are favourable for their proliferation (Harry, 1957, 1963; Wright, Anderson and McConachie, 1961).

Shell contamination can be minimized by general hygiene measures. The maintenance of the layers in colony cages with roll-away nests is an ideal arrangement. Roll-away nests should also be used with other systems of management. Where nest boxes are used, eggs should be removed as soon as possible after laying and a lining material should be used which has as low a microbial content as possible e.g. wood shavings (Harry, 1963) and this should be renewed at least 3 times a week or whenever it is seen to be soiled. The effectiveness of these measures can be conveniently assessed by periodically determining the degree of bacterial contamination on the shells of batches of eggs by means of the adhesive cellophane strip technique

(Thomas, 1961). This method was adapted for eggs at Coley Park, Reading. (M. McClintock, unpubl. data) where it was shown to give a reliable indication of the type of microorganisms present, though it could not be used quantitatively.

Some degree of shell contamination by the more commonly found microbes is however inevitable, and the shells of apparently clean eggs may contain more than 2×10^6 aerobic bacteria (Harry, 1963). Penetration of the shell is more likely to follow surface contamination with wet infected material. Williams, Dillard and Hall (1968) for example, found that moist faeces containing *Salm. typhimurium* when applied to the shells of eggs resulted in penetration of the cuticle and shell in 8·3% in 30 min/24° (75°F) and of the inner membranes of 11% in 2 h. No penetration was demonstrable however when dry faeces, similarly infected, was applied. For this reason soiled eggs, indicative of contact with wet faeces, should not be used for embryo or chick production.

Shell penetration is also assisted by fluctuations in egg temperature and for this reason eggs should be removed from communal nests as soon as possible. Studies made on the rate of penetration of *Pseudomonas* spp (Gordon and Tucker, 1954) have shown that the maximum amount of penetration occurs in the nest box when the egg cools. The majority of eggs are laid in the early part of the morning and eggs should be collected at

A B C

FIG. 4. Demonstration of the existence of relatively porous areas of the shell shown by immersing eggs in a dye solution. A, dye solution at egg temperature; B, dye solution refrigerated, and C, as (A) but under a pressure of 8 cm Hg.

least three times a day, twice in the morning. Eggs laid late in the day should not be allowed to remain in the nest overnight so as to be included in the collection made the following morning. After collection it is advisable to store the eggs at a constant temperature of *c.* 13° (56°F) and at a relative humidity of 70–80%.

Penetration of the shell occurs through relatively porous areas in the shell. These can be demonstrated by dye techniques as shown in Fig. 4. In cases where bacteria have penetrated the shell, areas of bacterial growth on the inner surface of the shell can be demonstrated by the triphenyltetrazolium chloride (TTC) method described by Board and Board (1967). Penetration is greatly facilitated by the presence of hair cracks which should be avoided by careful handling of the eggs.

On many farms it is now customary to disinfect the shell surface as soon as the eggs are collected, before the majority of the bacteria have had time to penetrate to sites inaccessible to the disinfectant. This procedure can be more effective than relying solely on the routine fumigation carried out subsequently at the hatchery. The most convenient method of disinfecting eggs on the farm is a 20 min period of fumigation with formaldehyde. Fumigation cabinets are now available for this purpose (Fig. 5). It is customary to generate the formaldehyde by the action of heat on paraformaldehyde, 1 oz. of this being used per 100 ft^3 of cabinet space. Alternatively formaldehyde may be generated from formalin ($4\frac{1}{2}$ fl. oz./100 ft^3) mixed with potassium permanganate in the ratio of 3:2 (Lancaster, Gordon and Harry, 1954). Care must be taken in individually designed cabinets and fumigation chambers, that adequate concentrations of gas are retained over the 20 min fumigation period as the success of the fumigation is related to the product of the gas concentration and the time of contact. A simple method of measuring terminal gas concentration is available which involves the use of a standardized solution of Schiff's reagent (Harry, 1959). In some cases a wet disinfection process has an advantage over fumigation, as when microbes are sealed within specks of dirt or when it is desired to use a disinfectant particularly active against certain contaminants. Sodium pentachlorophenate 0·5%, chlorine or iodine compounds (250–1000 ppm of active ingredient), cetyl trimethyl ammonium bromide (CTAB) 1%, or benzalkonium chloride 0·1% have been found suitable as general disinfectants, and phenylmercury dinaphthylmethane disulphonate (PMDD) 0·2% has been found suitable for disinfecting shells contaminated with the spores of *A. fumigatus* (Harry and Cooper, 1970). Wet disinfection processes however require a more skilled operator. The temperature of the dip solution has to be carefully controlled and the solutions require renewal before they are inactivated. Neglect of these aspects will increase rather than reduce the contamination of the eggs.

Eggs may also be treated to control a congenital infection with *Mycoplasma gallisepticum*. For this purpose, eggs, after surface disinfection, are immersed in a refrigerated solution of an anti-mycoplasmal drug. Alternatively the drug may be inoculated through the shell. Details of these and other methods for the elimination of *M. gallisepticum* are described by Osbaldiston and Wise (1968) and Yoder (1970).

FIG. 5. Fumigation cabinet suitable for fumigating eggs on farm or small hatchery.

Sources of Chick Infection in the Hatchery

Contamination by fluff

Chick fluff consists of down spicules which are coated with the dried embryonic fluids. Bacteria multiply rapidly in these fluids before the chicks dry after hatching and counts of up to 30×10^6 bacteria g fluff are often encountered. The dried fluff is sufficiently light to travel on air currents and can also contaminate floors. In the hatcher it contaminates the air inhaled by the chicks and penetrates the perforations of the shell made by late hatching chicks.

Unless prevented, it can drift from one hatcher to another and from the hatching room to other parts of the hatchery on air currents or on the footwear or clothes of the staff. In this way it can contaminate eggs as yet unhatched and form a source from which subsequent batches of chicks can be infected. The spread of infection by contaminated fluff can be reduced by:

1. connecting the ventilators of the hatcher to the outside of the building and also by accommodating the hatchers and setters in separate rooms;
2. adjusting the ventilation of the hatchery so that the air from the hatching room does not flow into rooms containing eggs at an earlier stage of development;
3. restricting the movement of the staff from one section of the hatchery to another;
4. washing the floors frequently with a disinfectant solution;
5. liberation of a volatile disinfectant in an aerosol form during the hatching period to disinfect the smaller sized airborne fluff particles;
6. transfer of fluff and debris from the hatching trays to disposable containers which are subsequently sealed;
7. removal of fluff deposits from emptied hatchers by vacuum cleaners or by flushing into floor drains with a hose, and
8. cleaning and disinfecting trays and incubators between each hatch.

The effectiveness of these and the other measures of disease control mentioned can be determined by:

1. monitoring the air in different parts of the hatchery using a sieve sampler (Anderson, 1958), a slit sampler (Bourdillon, Lidwell and Thomas, 1941) or gravity plates—for additional details, see p. 37. Whatever sampling system is used it is necessary for it to be carried out at regular intervals over a period of several months in order that a reliable assessment can be made of changes in the hygiene status of the hatchery;
2. by assessing the microbial content of the fluff (Nichols, Leaver and Panes, 1967), and
3. by measuring the amount of surface contamination on equipment and incubators, etc., by swab culture, and adhesive cellophane method (Thomas, 1961) or by the "agar sausage" technique (ten Cate, 1965).

Contamination within the incubator

Embryos may become infected by contact with infected fluids or mould spores which may escape if a decomposing egg breaks in the incubator. This occurrence should be rare however if the recommended standard of hygiene and care in handling has been maintained on the supply farm. Infection can also result if the water containers associated with the incubator humidifiers are not regularly cleaned. M. McClintock (unpubl. data) noted an instance where pseudomonads had become established in the storage tank water which resulted in the contents of the incubator being contaminated whenever the humidifiers were brought into operation.

Contamination by handling

Disease agents can also be brought into the hatchery by the hatchery staff. It is customary to ensure that hatchery staff have no outside contact with poultry, but apart from specific poultry pathogens there is a risk of bacteria of human origin such as salmonellae or *Staph. aureus* being introduced in this way. Transmission of infection occurs when eggs or chicks are handled and for this reason it is advisable for hands to be disinfected by the application of disinfecting creams or lotions. Preparations suitable for the purpose have been reviewed by Lowbury, Lilly and Bull (1963). These should be reapplied when handling chicks from different farms of origin.

The main aspects of hatchery hygiene have been discussed but further details on these and related hygiene measures are to be found in the review articles by Lancaster (1961, 1970), Harry and Gordon (1966) and relevant publications of the Ministry of Agriculture Fisheries and Food (1964, 1966).

References

ANDERSON, A. A. (1958). New sampler for the collection, sizing and enumeration of viable airborne particles. *J. Bact.*, **76**, 471.

BIESTER, H. E. & SCHWARTE, L. H. (1965). *Diseases of Poultry*. Ames, Iowa: State Coll. Press.

BIGGS, P. M. (1970*a*). Production of pathogen free avian cell substrate for production of vaccines. *Lab. Pract.*, **19**, 45.

BIGGS, P. M. (1970*b*). Marek's disease – Prospect for Control. 14*th Wld's Poult. Congr. Madrid.*

BOARD, P. A. & BOARD, R. G. (1967). A method of studying bacterial penetration of the shell of the hen's egg. *Lab. Pract.*, **16**, 471.

BOURDILLON, R. B., LIDWELL, O. M. & THOMAS, J. C. (1941). A slit sampler for collecting and counting airborne bacteria. *J. Hyg., Camb.*, **41**, 197.

BURMESTER, B. R. (1962). The vertical and horizontal transmission of avian visceral lymphomatosis. *Cold Spring Harb. Symp. quant. Biol.*, **27**, 471.

BUXTON, A. (1957). *Salmonellosis in Animals—a Review*. Commonwealth Agric. Bur., Farnham Royal, Bucks.

BUXTON, A. & GORDON, R. F. (1947). The epidemiology and control of *Salmonella thompson* infection of fowls. *J. Hyg., Camb.*, **45**, 265.

TEN CATE, L. (1965). A Note on a simple and rapid method of bacteriological sampling by means of agar sausages. *J. appl. Bact.*, **28**, 221.

CHUBB, R. C. & CHURCHILL, A. E. (1968). Precipitating antibodies associated with Marek's disease. *Vet. Rec.*, **83**, 4.

CHUTE, H. L., STAUFFER, D. R. & O'MEARA, D. C. (1964). The production of specific pathogen free broilers in Maine. *Bull. Me agric. Exp. Stn.*, 633.

COOPER, D. M. (1970). Poultry: principles of disease control. *Vet. Rec.*, **86**, 388.

EGAN, B. J. & BUTLER, E. J. (1972*a*). Controlled environment systems for experimental animals. Part I. A unidirectional air flow brooder for chicks. *Lab. Anim.*, **6**, 23.

EGAN, B. J. & BUTLER, E. J. (1972*b*). Controlled environment systems for experimental animals. Part II. A unidirectional air flow tent. *Lab. Anim.*, (in press).

FABRICANT, J. & LEVINE, P. P. (1951). Studies on the diagnosis of Newcastle disease and Infectious Bronchitis. IV. The use of the serum neutralisation test in the diagnosis of Infectious Bronchitis. *Cornell Vet.*, **41**, 68.

FABRICANT, J., LEVINE, P. P., CALNEK, B. W., ADLER, H. E. & BERG, J. R. (1959). Studies of egg transmission of PPLO in Chickens. *Avian Dis.*, **31**, 197.

FITCH, C. P., LUBBEHUSEN, R. E. & DIKMANS, R. N. (1924). Report of experimental work to determine whether avian tuberculosis is transmitted through eggs of tuberculous fowls. *J. Am. vet. med. Ass.*, **66**, 43.

GORDON, R. F. (1942). Practical application of the rapid whole blood test for pullorum disease. *Vet. Rec.*, **54**, 495.

GORDON, R. F. & TUCKER, J. F. (1954). Behaviour of *Pseudomonas* spp. and the natural occurrence of the organism in the fowl and its environment. 10*th Wld's Poult. Congr.*, 348.

GORDON, R. F. & TUCKER, J. F. (1956). Unpublished findings.

GRUMBLES, L. C., HANSON, R. P., ROSENWALD, A. H., VAN ROEKEL, H., BIRINS, J. A. & HEJL, J. M. (1963). *Methods for the examination of poultry biologics*. 2nd Ed. Nat. Acad. Sci. Washington D.C: Nat. Res. Council.

HARRY, E. G. (1957). The effect on embryonic and chick mortality of yolk contamination with bacteria from the hen. *Vet. Rec.*, **69**, 1433.

HARRY, E. G. (1959). A method of estimating the formaldehyde vapour concentrations used in egg fumigation. *Vet. Rec.*, **71**, 842.

HARRY, E. G. (1963). The relationship between egg spoilage and the environment of the egg when laid. *Br. Poult. Sci.*, **4**, 91.

HARRY, E. G. & COOPER, D. M. (1970). The treatment of hatching eggs for the control of egg transmitted Aspergillosis. *Br. Poult. Sci.*, **11**, 273.

HARRY, E. G. & GORDON, R. F. (1966). Egg and hatchery hygiene. *Veterinarian*, **4**, 5.

HARRY, E. G. & HEMSLEY, L. A. (1965). The relationship between environmental contamination with septicaemia strains of *E. coli* and their incidence in chickens. *Vet. Rec.*, **77**, 241.

HUNGERFORD, T. G. (1968). A clinical note on avian cholera. *Aust. vet. J.*, **44**, 31.

JORDAN, F. T. W. (1956). Some observations on respiratory diseases in poultry. *Vet. Rec.*, **68**, 554.

LANCASTER, J. E. (1961; 1970). *Hatchery Sanitation—a Review.* Canada: Health of Animals Division. Dept. of Agriculture.

LANCASTER, J. E., GORDON, R. F. & HARRY, E. G. (1954). Studies on disinfection of eggs and incubators. III. The use of formaldehyde at room temperature for the fumigation of eggs prior to incubation. *Br. vet. J.*, **110,** 238.

LEACH, R. H. & BLAXLAND, J. D. (1966). The need for the standardisation of serological techniques for the detection of *Mycoplasma gallisepticum* infection in poultry. *Vet. Rec.*, **79,** 308.

LOWBURY, E. J. L., LILLY, H. A. & BULL, J. P. (1963). Disinfection of hands: removal of resident bacteria. *Br. med. J.*, **1,** 1251.

MINER, R. W. (1952). The chick embryo in biological research. *Ann N.Y. Acad. Sci.*, **55,** 37.

Ministry of Agriculture, Fisheries & Food (1964). *Incubation and Hatchery Practice.* Bull. 148, London: H.M.S.O.

Ministry of Agriculture, Fisheries & Food (1966). *Poultry Health Scheme Regulations.* Tolworth, Surrey: Animal Health Division.

NEW, D. A. T. (1966). *The culture of vertebrate embryos.* London: Logos Press and Academic Press.

NICHOLS, A. A., LEAVER, C. W. & PANES, J. J. (1967). Hatchery hygiene evaluation as measured by microbiological examination of samples of fluff. *Br. Poult. Sci.*, **8,** 297.

OSBALDISTON, C. W. & WISE, D. R. (1968). Methods for the eradication of *Mycoplasma gallisepticum. Poult. Rev.*, **7,** 75.

PAYNE, L. N. (1967). *The Chick Embryo.* In *Handbook on the care and management of laboratory animals.* 3rd ed. p. 777. Edinburgh and London: UFAW. Livingstone.

PAYNE, L. N. (1968). "Eggs in virology". In *Egg quality—a study of the hen's egg.* (T. C. Carter, ed.) p. 181. Edinburgh: Oliver & Boyd.

RETTGER, L. E. & PLASTRIDGE, W. N. (1932). Variants of *Salmonella pullorum. J. infect. Dis.*, **50,** 146.

RUBIN, H., CORNELIUS, A. & FANSHIER, L. (1961). The pattern of congenital transmission of an avian leukosis virus. *Proc. natn. Acad. Sci. U.S.A.*, **47,** 1058.

TAYLOR, L. W., LOWRY, D. C. & RAGGI, L. G. (1955). Effects of an outbreak of Avian Encephalomyelitis (Epidemic Tremor) in a breeding flock. *Poult. Sci.*, **34,** 1036.

THOMAS, M. (1961). The sticky film method of detecting skin staphylococci. *Mon. Bull. Minst. Hlth.*, **20,** 37.

VAN ROEKEL, H. (1946). *Proc. Conf. Newcastle disease* p. 17. U.S. Dept. Agric.

WILLIAMS, J. E., DILLARD, L. H. & HALL, G. O. (1968). The penetration patterns of *Salmonella typhimurium* through the outer structures of chicken eggs. *Avian Dis.*, **12,** 445.

WRIGHT, M. L., ANDERSON, G. W. & MCCONACHIE, J. D. (1961). Transmission of aspergillosis during incubation. *Poult. Sci.*, **40,** 727.

YODER, H. W. (1970). Preincubation heat treatment of chicken hatching eggs to inactivate mycoplasma. *Avian Dis.*, **14,** 75.

ZARGER, S. L. & POMEROY, B. S. (1950). Isolation of Newcastle disease virus from commercial fowl pox and laryngotracheitis vaccines. *J. Am. vet. med. Ass.*, **116,** 304.

Methods of Handling and Testing Starter Cultures

W. A. Cox

Unigate Central Laboratory, Western Avenue, Acton, London W.3, England

AND

J. E. Lewis

Unigate Foods Ltd., Bailey Gate, Sturminster Marshall, Wimborne, Dorset, England

In the manufacture of fermented milk products there are certain essential requirements for the starter cultures which are used. These may be summarized as follows:

1. Cultures should be capable of forming lactic acid with or without flavour compounds in milk at a suitable rate for the type of cheese or fermented product being made.
2. Cultures should be free of contamination by: extraneous yeasts and moulds, coliforms, bacteriophage (except in specific bacteriophage carrying cultures), and other microbial contaminants at levels liable to interfere with the quality of the product.

Classification of Starter Cultures

(*Excluding continental cheese types*)

Cheese starter cultures

Because of the susceptibility of lactic streptococci to bacteriophage attack a number of different approaches to avoiding failures due to this cause have been developed.

Single strain cultures

The use of single strain cultures where adopted is associated with the aseptic transfer and growth of all cultures including bulk starter. Vat ripening of milk is restricted to less than 30 min. Carefully selected rotations of bacteriophage unrelated cultures are used extending over at least 6 days. This method had been adopted in Australia and New Zealand.

Paired compatible strains

These may be used in a similar way to single strains. Aseptic precautions are necessary for all culture stages, and a rotation of strains is desirable. This procedure is used extensively in the United States.

Commercial mixed strains

These cultures are most reliable when grown under aseptic conditions up to and including the bulk starter stage. Mixed strain cultures are generally used in this country for Cheddar and Territorial cheese manufacture. They are regarded as less prone to failure where the longer vat ripening periods (30–70 min) are used for Cheddar making.

Bacteriophage carrying starter cultures

These types of culture have been fairly widely adopted in the Netherlands. Careful control and replacement from a central laboratory are apparently desirable.

Starter cultures for hard cheese

The starter cultures used for the manufacture of Cheddar and Territorial varieties of English hard pressed cheese may be either single or mixed strains of lactic streptococci.

Single strains	*Streptococcus lactis* *Streptococcus cremoris* *Streptococcus diacetilactis* *Streptococcus lactis* var *diacetilactis*; Buchanan, Holt and Lessel (1966).
Mixed strains	Any combination of strains of the above species with or without *Leuconostoc* spp

Single strains are generally paired for the vat ripening of milk during cheese manufacture.

Starter cultures for soft cheese

Full fat and medium fat soft cheese are manufactured with mixed strain cultures containing *Str. diacetilactis* or *Leuconostoc* strains.

Starter cultures for cottage cheese

Cottage cheese manufacture may be undertaken with paired strain or mixed strain cultures of *Str. lactis* or *Str. cremoris*. It is necessary to avoid using cultures which may contain strains of *Str. diacetilactis* and *Leuconostoc* spp capable of causing "floating curds". The ripening of cream for cottage cheese may be undertaken with strains of *Str. diacetilactis* or mixed cultures containing this species.

Starter cultures for other fermented milk products

Cultured buttermilk and soured cream are manufactured from mixed starter cultures containing *Str. diacetilactis* strains capable of high level of diacetyl and other flavour compound production. Yogurt is manufactured using starter cultures of mixed species of *Lactobacillus bulgaricus* and *Str. thermophilus*. The species may be maintained separately and mixed prior to the preparation of mother cultures or the mixture may be propagated in the laboratory.

Methods of Starter Culture Propagation

Laboratory transfer method

The traditional method of preparing cultures (mother cultures) for cheese-making is based on the daily subculture of starter cultures in pasteurized (or sterilized) milk in test tubes. In addition, these initial laboratory cultures are inoculated into sterilized or pasteurized milk in glass or polypropylene bottles (intermediate or working cultures). Alternatively, subculture of the intermediate cultures daily, omitting the initial culture stage, may be practised. With both systems, the intermediate cultures are inoculated into bulk cultures which are used subsequently for vat inoculation.

Marschall Laboratories method

This is a commercially available system based on the growth of selected strains in a special milk based medium (Marstar). The advantage claimed for this system is that bacteriophage multiplication will not occur in the growth medium with the selected strains supplied. These cultures may be grown through the normal sequence of mother culture, working culture and bulk culture, or alternatively, concentrated cultures for direct bulk medium inoculation may be used.

Hansens Laboratories method

Preparations of concentrated starters for bulk inoculation have recently been made available.

Lewis protected method

This method, which was originally described by Lewis (1956) is based on the complete aseptic handling of starter cultures from the master culture (freeze dried or liquid culture), through the intermediate stages of mother and working culture up to and including, the bulk starter. Since the method of protection depends on the exclusion of airborne "phage" and other contaminants, maintaining sterility of the apparatus and correct handling of the equipment is essential. With close supervision this method is successful in achieving its objective in the hands of trained operators. For the method to be universally accepted, the apparatus must be of high quality to ensure ease in operation and must incorporate safeguards against misuse. The technique for transferring cultures is essentially similar to that described originally (Lewis, 1956) but specific improvements that have been made are detailed below.

Needle assemblies

Originally the double ended needle units were manufactured from blood transfusion cocks and 18 gauge stainless steel needles with bosses threaded to retain stainless steel discs. These discs assisted in handling the assemblies and acted as stops against the seals. The assemblies were held together by pressure only and it was possible for them to fall apart if handled badly. As higher solids content milk was used, 16 and then 14 gauge needles were substituted without causing leakage or damaging the seals. Various models of needle assemblies manufactured in one piece are now available (R. G. Hooper Ltd., 69 Leigh Road, Wimborne, Dorset; Carsberg and Froud Ltd., Rayne Road, Braintree, Essex).

The latest model, fitted with a 12 gauge needle, is shown in Fig 2. It incorporates a water sealed hood which protects the tap, which is empty during boiling but filled with hypochlorite solution during inoculation and sampling.

Large needle assemblies for bulk starter inoculation are made of the same design (Fig. 12). These assemblies are made to ensure complete heat transfer to all parts during sterilization; they can also be broken down for examination and replacement of parts.

Plastic bottles

Polythene bottles of the 4 oz. mother culture size are satisfactory and have not been changed. They are virtually indestructible in the conditions under which they are used and remain satisfactory for years. The large bottles were originally made up of two parts, consisting of standard bottles of 20 or 40 oz. capacity with a separate specially moulded neck piece welded on to take the seal. These moulded necks were designed internally to hold the special seal in use. The rough inner surface of the weld caused difficulties in cleaning and sterilizing, and the push-on cap tended to pull off in handling. A new bottle of 30 oz. total capacity has been produced fully blown in one piece with an external screw thread for the cap and is now the standard bottle incorporating all the features found necessary.

Seals

Seals designed to fit $\frac{1}{2}$ in. reductase tubes and bottles of similar neck dimensions and to vent under pressure during sterilization were used originally. By arranging a very tight fit in the neck of the plastic bottles, these seals withstood the pressure increase during heat treatment of the milk but, due to faults in manufacture, leaks sometimes occurred which in this aseptic system could not be tolerated. These seals are now being replaced by an alternative closure, specifically designed to withstand pressure and hold a water seal during inoculation.

Milk media

It is often difficult to understand the reasons for the failure of starters, which are affected by the conditions in which they are stored or handled, the apparatus used and the numerous factors which bear on cheese production. These include the milk or media which is used for cheesemaking and starter production, and while the cheese milk cannot be altered to any great extent, it is possible to control the milk used for laboratory cultures and bulk starter preparation. It is now fairly standard practice to use antibiotic—free skim milk powder produced under the best seasonal conditions and to maintain the same supply for 6–12 months. This ensures that depots continually receive the same cultures in the same condition just prior to their own bulk starter preparation.

Starter handling procedure

The handling of starter cultures using the Lewis (1956) system is clearly illustrated in Figs 1–10. From the master culture stage (Fig. 1), in which the freeze dried culture has been grown in pasteurized milk in a medical flat bottle, clotted culture is transferred to 4 oz. mother culture. Using the technique illustrated in Fig. 2 mother culture may be transferred regularly (Fig. 3). Samples taken for laboratory tests (Fig. 4) may be examined for contamination, activity or vitality.

Working cultures (30 oz.) of pasteurized milk may be inoculated from

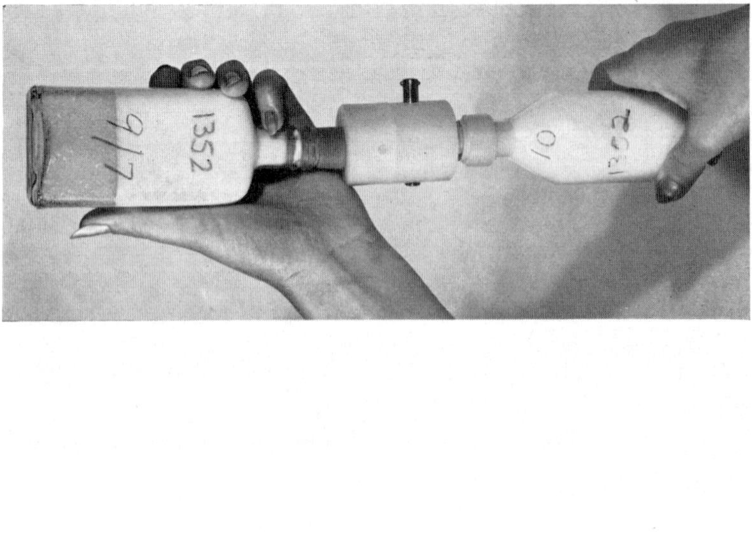

FIG. 1. Freeze dried master; clotted in 24 h after awakening. Inoculation into first mother culture.

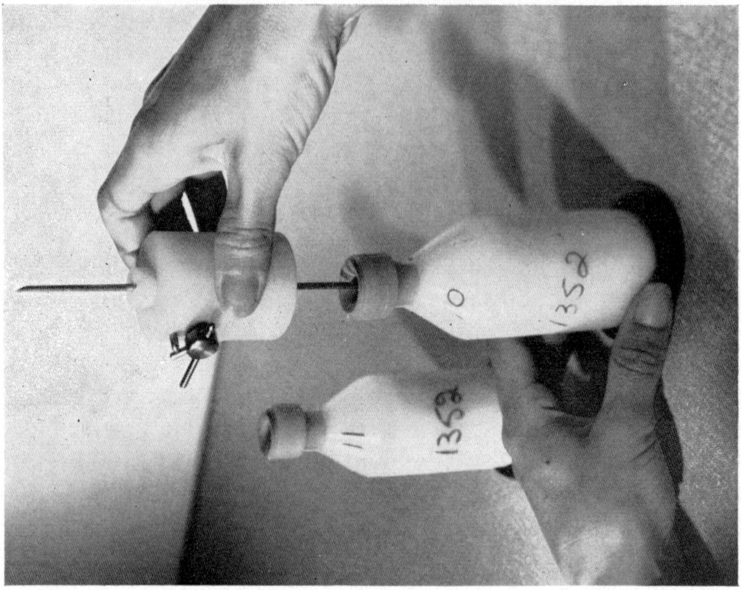

FIG. 2. Inserting water sealed inoculating needle through water seal into first clotted mother culture.

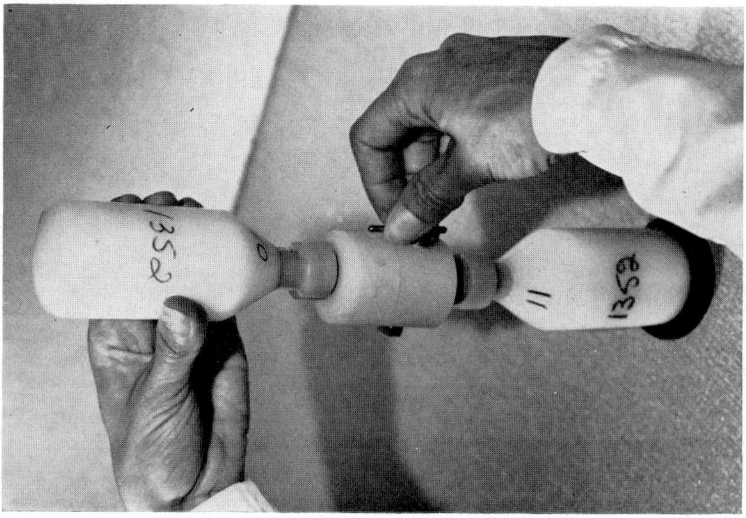

FIG. 3. Daily propagation. Inoculating second mother culture 4 oz. containing prepared milk media from first clotted mother culture.

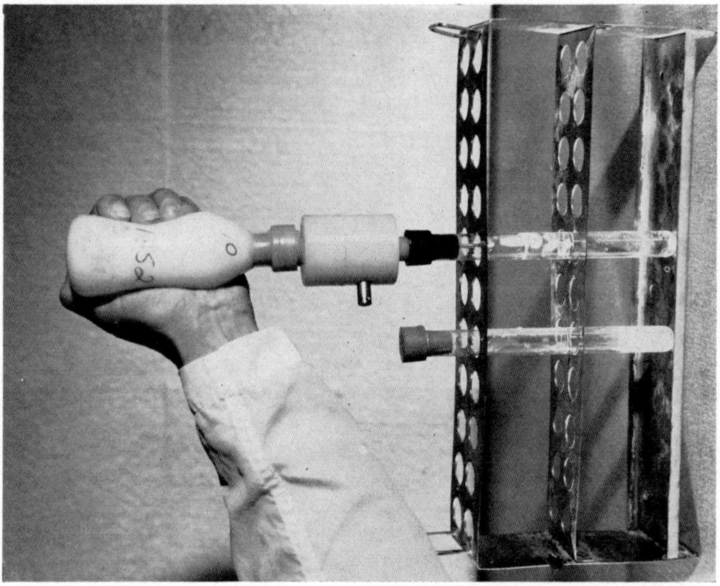

FIG. 4. Sampling mother culture for laboratory tests.

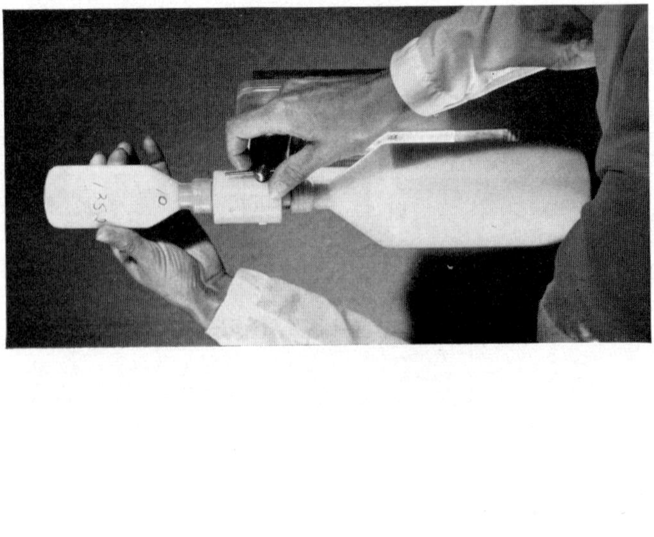

FIG. 5. Inoculating working culture for batch or quantity production.

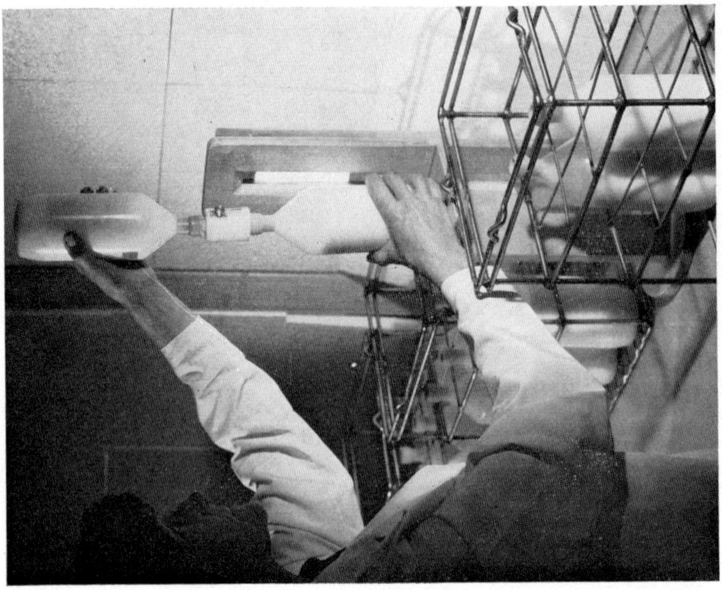

FIG. 6. Quantity production. A single 30 oz. working culture for inoculating 50 batch working cultures.

FIG. 7. Incubation of mother culture and working cultures. 18 h at 22°.

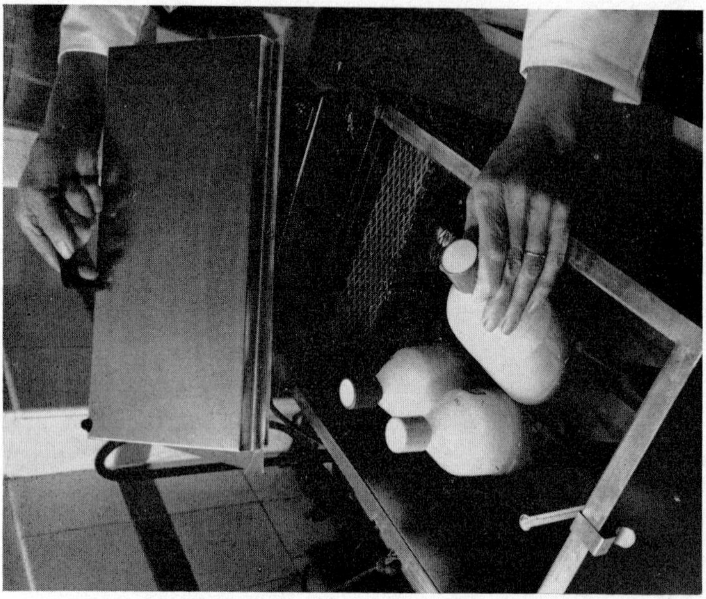

FIG. 9. Preparation of deep frozen working cultures for bulk starter inoculation. Two hours at 30° (water bath) followed by 18 h at 22° (incubator).

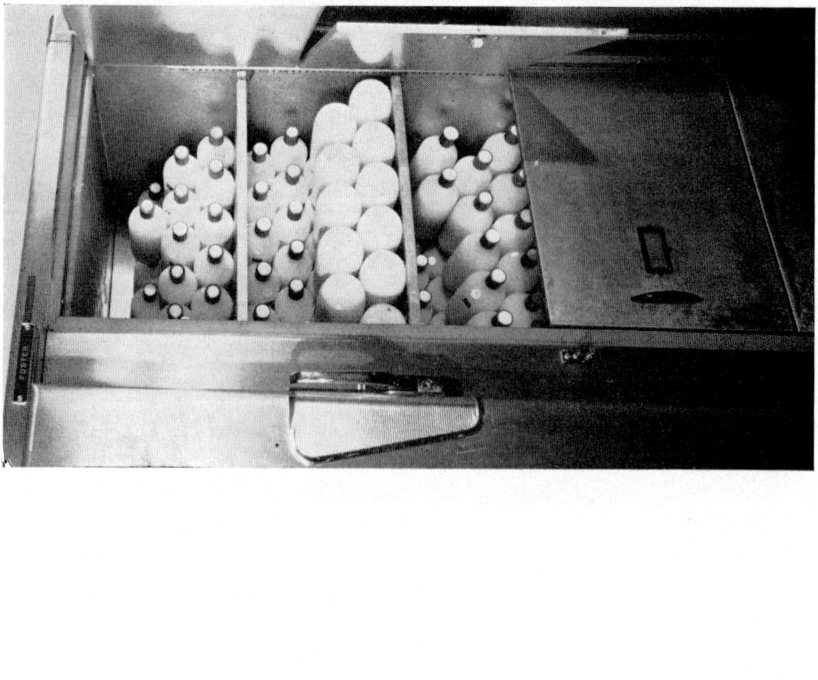

FIG. 8 (*left*). Deep frozen storage of working cultures. Within one hour to 0°, and 16 h to −20° C. Held 3–6 months −20°C.

4 oz. mother cultures (Fig. 5) or batches of working cultures inoculated from 30 oz. clotted working cultures (Fig. 6). All mother cultures and working cultures are incubated for 18 h at 22° (Fig. 7) for propagation purposes. Working cultures for storage and distribution are deep frozen after inoculation (Fig. 8).

The preparation of working cultures for bulk starter inoculation is effected by thawing deep frozen cultures for 2 h at 30° (Fig. 9), followed by incubation at 22° for 18 h.

Bulk starter vessels

Pressure containers (10 and 40 gallon capacity) operating under completely sealed conditions have been standard equipment under this system. Inoculation of these is shown in Figs 10 and 11. For modern cheese creameries, the need for large, automatically controlled pressure vessels has arisen, and 250–500 gallon size vessels have been developed for bulk starter culture and the inoculation of a 250 gallon vessel is illustrated in Fig. 12. These have been designed to work under fully protected conditions and are considered to be an advance on other types of large vessels of various capacities which depend on filtered air. The complications arising from the latter type of arrangement make full protection difficult to achieve.

Large scale working starter production

It is essential to provide facilities for washing and sterilizing considerable quantities of bottles, needles, seals and the equipment in use. With a large demand for starter cultures for direct inoculation into bulk starter at cheese manufacturing creameries, storage space is necessary to hold considerable stocks. After preliminary trials, deep frozen conditions (—20°C) were found to be suitable and large stocks of cultures can be stored in deep freeze cabinets. The advantages lies in being able to accumulate and hold a variety of cultures in order to provide selections or rotations and to offer alternative cultures at a moment's notice. Starters can be produced on a batch system which enables better check controls to be carried out in the laboratory and cheese room before despatch. Suitable arrangements have been made for transporting the 30 oz. bottles in insulated containers which will hold them in frozen conditions for 72 h.

FIG. 10. Inoculation. Ten gallon bulk starter vessel fitted with water sealed lid.

FIG. 11. Inoculation. Forty gallon vessel with water sealed lid.

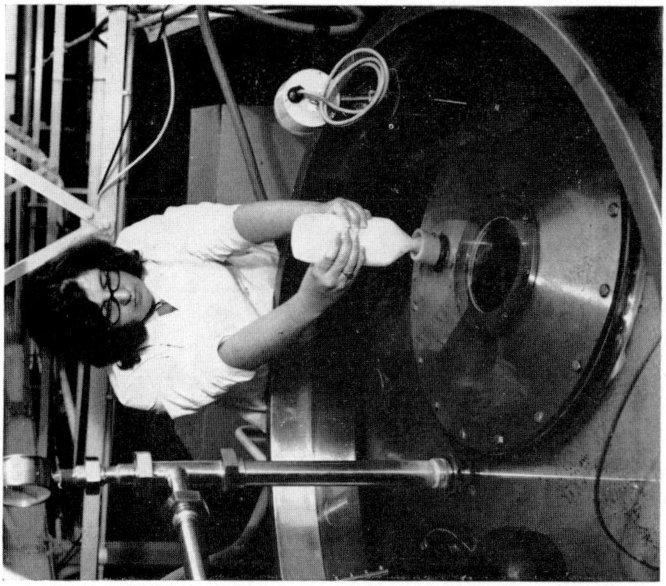

FIG. 12. Inoculation of 250 gallon vessel through depth of water contained in the jacket.

Methods of Evaluating the Suitability of Starter Cultures for Cheese Manufacture

Cheesemaking

The most important assessment for the acceptability of starter cultures for cheesemaking is that the product should be of high quality, without "off" flavours, and made in normal making times. An experimental vat assessment is necessary before the general use of new lactic cultures.

Vitality test

It is possible to simulate in part, the cheesemaking process on the laboratory scale. The method first described by Whitehead and Cox (1932) for the assessment of the suitability of starter cultures for cheesemaking has been modified in current practice (Pulay and Zsmko, 1969). This type of test has some value at the present time for assessing the usefulness of commercial mixed starters for Territorial cheese varieties where rapid acid development is necessary during manufacture.

A method of vitality testing which is used at the present time for assessing cultures for Cheshire and Cheddar manufacture is detailed below.

Test procedure

Place 500 ml vat milk in stoppered bottle in 30° water bath. After 15 min carry out the following schedule.

0.0 h	Inoculate with 10 ml starter and mix.
1.30 h	Dilute 8 drops Standard Rennet in 5 ml Ringer's solution, add all to the milk, and mix.
2.00 h	Shake bottle 10 times.
3.00 h	Pour off about 30 ml whey for acidity. Pour curds and whey into a sieve to drain and shake occasionally, while doing acidity. When only a thin trickle of whey is coming through, tip curds from sieve into a 400 ml beaker, place in a 6 × 6 in. wire basket in 30° bath. The basket is to prevent the beaker floating and tipping over.
4.00 h	Remove 10 ml whey for acidity and sieve again. The acidity is always measured before sieving in case of whey loss.
5.00 h 6.00 h	as 4.00 h.

Acidity determination. Titration procedure used is identical with that described under activity tests.

Interpretation. Suitably active cultures should show an approximate doubling of acidity between the 5th and 6th h, with a final acidity at 6 h $>0{\cdot}5\%$ lactic acid.

Activity test

A simple laboratory test first described by Anderson and Meanwell (1942) gives a measure of the rate of acid formation by starter cultures in milk. A 1% of starter inoculated into pasteurized cheese milk and incubated in duplicate tubes for 6 h at 30°. The acidity is determined by titration of 10 ml culture with N/9-NaOH using 1 ml 0·5% phenolphthalein. The figure obtained is referred to as the **Activity**. This test is much simpler than the Vitality Test and can be regarded as giving less information in terms of starter potential during cheesemaking. From experience with mixed starters, cultures of very low activity would be considered potentially unsatisfactory, whereas cultures of high activity would not necessarily be more satisfactory than those of medium activity.

Methods of Detecting Contaminants in Starter Cultures

After incubation of mother, working or bulk cultures for 20–24 h at 22°, the following tests are undertaken (Anon, 1970).

Yeasts and moulds

Samples transferred by loop from incubated mother, working or bulk cultures on to Malt Agar and streak plates prepared. The plates are incubated at 26° for 3–5 days and examined for growth. Cultures should be free of yeast and mould contamination.

Coliforms (presumptive)

One loopful is transferred from incubated culture into single strength MacConkey Broth. Broth cultures are incubated at 30° for 3 days. Positive results reported where acid and gas production is recorded.

Quantitative estimation of gas forming sporebearers

Where pasteurized milk (full cream, skim or reconstituted powder) is used for starter culture, the incidence of gas-forming sporebearers should be checked on representative mother or working cultures prepared in each batch of milk. A suitable method is as follows: 10 ml, 1 ml and 0·1 ml of culture are transferred in triplicate to 10 ml quantities of Bromocresol purple milk. The tubes are heated at 80° for 10 min and then overlaid with 3 ml of 2% (w/v) agar solution. All tubes transferred to 37° water bath and examined for gas formation after 7 days' incubation. Results are expressed as the Most Probable Number of gas forming sporebearers per 100 ml culture using an MPN Table (Jacobs and Gerstein, 1960). Numbers of sporebearers in excess of 1800/100 ml milk would be regarded as unsatisfactory and requiring further investigation.

Bacteriophage Detection Test (PDT) Applied to Starter Cultures

Bacteriophage contamination of starter cultures may occur in dairy environments and on subsequent incubation of the cultures, lysis of the starter strains could lead to serious loss of activity. It is important to check that cultures are free from bacteriophage except when specific phage carrying cultures are being prepared.

General method (Anderson and Meanwell, 1942)

Reconstituted skim milk containing 1% of starter culture is transferred aseptically to the required number of tubes. One tube is reserved as a control for the test. Suitably diluted samples of working culture and bulk starter prepared in quarter strength Ringer's are added to the appropriate tubes containing homologous mother culture. The tubes are incubated for 6 h at 30° and then titrated for acid development. If the tubes containing diluted samples of working or bulk starter show a significantly lower acidity compared with the control mother culture, those cultures may be assumed to be infected with bacteriophage. The mother cultures should also be tested against master cultures for phage contamination. The examination of cheese room whey is a useful guide to the occurrence of bacteriophage affecting particular starter cultures in creamery use.

Simplified method

By the use of Bromocresol Purple (BCP) Milk it is possible to simplify the bacteriophage detection test. The direct acidification of milk is wholly or partially inhibited by bacteriophage; therefore, after incubation, the yellow acid control cultures may be contrasted with blue or green inhibited test cultures.

Plaque methods

It is possible to show conclusively the presence of bacteriophage by plaque formation on suitably prepared agar media (Meanwell and Thompson, 1959). This procedure is not normal creamery practice but is valuable for investigation purposes particularly where mixed cultures are being studied. Commercial mixed cultures often show no reaction to different phage sources when both the BCP milk and plaque methods are used. Strains isolated from the mixed cultures must be tested against suspected phage containing material before confirming the absence of phage.

Fermented Milk Products

Yogurt, butter milk, sour cream

Mother, working and bulk cultures may be prepared by a similar technique to that used for preparation of cheese starter cultures. Deep frozen working cultures may also be prepared for storage and distribution.

Methods for detection of contaminants

Yeasts and moulds

The methods of examination for yeasts and moulds are similar to those employed in checking cheese starters modified for yogurt by the further overnight incubation of clotted working cultures at 22° before testing.

Coliforms

The tests are the same as those for cheese starters.

Suitability for product manufacture

Yogurt

An activity test similar in principle to the cheese culture test is used but based on incubation for 4–4½ h at 42°.

Buttermilk, sour cream

Mixed cultures of lactic streptococci capable of developing acidity and aroma in the product are used. Aroma tests can be carried out by the determination of diacetyl and acetaldehyde but it is important that a taste test of the product is also made (Mocquot and Hurel, 1970).

Acknowledgement

We would like to thank Mr. E. Merry for the excellent photographs of different stages of starter propagation.

References

ANDERSON, E. B. & MEANWELL, L. J. (1942). The problem of bacteriophage in cheesemaking. I. Observations and investigations on slow acid production. *J. Dairy Res.*, **13,** 58.

ANON (1970). Supplement I (1970) to BS 4285: 1968. Methods of Microbiological Examination of Milk Products. Paras. 3.5., 4.4./3.5.5. London: B.S.I.

BUCHANAN, R. E., HOLT, R. G. & LESSEL, E. F. JR. (eds) (1966). *Index Bergeyana: an annotated alphabetic listing of names of the taxa of the bacteria*. Edinburgh, London: E. & S. Livingstone Ltd.

JACOBS, M. B. & GERSTEIN, M. J. (1960), *Handbook of Microbiology*. New York: Van Nostrad.

LEWIS, J. E. (1956). A new approach to the problem of phage control during the production of commercial cheese starters. *J. Soc. Dairy Technol.*, **9,** 123.

MEANWELL, L. J. & THOMPSON, N. (1959). The influence of rennet on bacteriophage infection in the cheese vat. *J. appl. Bact.*, **22,** 281.

MOCQUOT, G. & HUREL, C. (1970). The selection and use of some microorganisms for the manufacture of fermented and acidified milk products. *J. Soc. Dairy Technol.*, **23,** 130.

PULAY, G. & ZSMKO, M. (1969). Activity test of mesophilic starters. *Tejepar*, **18,** 25.

WHITEHEAD, H. R. & COX, G. A. (1932). A method for the determination of vitality in starters. *N.Z. J. Sci. Technol.*, **13,** 304.

Actinomycete and Fungus Spores in Air as Respiratory Allergens

J. LACEY

Rothamsted Experimental Station, Harpenden, Hertfordshire, England

J. PEPYS

Institute of Diseases of the Chest, Brompton, London S.W.3, England

AND

T. CROSS

Postgraduate School of Studies in Biological Sciences, University of Bradford, Bradford, Yorkshire BD7 1DP, England

The air almost always contains spores, but their numbers and types depend on the time of day, weather, season and geographical location (Gregory, 1961). Usually spores cause no trouble to most people, but they can be harmful by provoking allergic responses or causing infections, or because they contain toxins. Here we are concerned mostly with spores that cause respiratory allergy, and especially with those encountered in agricultural or industrial buildings or laboratories, where concentrations exceed those usual in outdoor air. Spore concentrations outdoors seldom exceed $10^5/m^3$, and only rarely reach $10^6/m^3$ (Gregory and Sreeramulu, 1958). Indoors, spore concentrations vary widely depending on local sources and concentrations may be large in working environments. For instance, in farm buildings up to $2{\cdot}9 \times 10^9$ spores/m^3 have been found while mouldy fodder was being handled (Lacey and Lacey, 1964; Lacey, 1969, 1971*b*).

Exposure to Spores in Working Environments

Agriculture

Exposure to dust from mouldy hay and grain has been shown to produce severe respiratory symptoms in some agricultural workers. Symptoms resembling Farmer's Lung were first described more than 200 years ago (Ramazzini, 1713), but only recently have the causative agents been identified. The background to Farmer's Lung disease provides much useful information for detecting or anticipating similar diseases in other environments.

Damp hay in a stack or bale provides an environment in which bacteria and fungi grow rapidly, so providing the heat that favours the growth of thermophilic actinomycetes. Spores of the two species chiefly responsible for Farmer's Lung, *Micropolyspora faeni* and *Thermoactinomyces vulgaris*, contaminate hay from the soil. They germinate and the hyphae ramify through the hay producing many thousands of spores, up to 10^{10}/g of hay. Such samples of mouldy hay also contain many other spores than of actinomycetes, and release clouds of spores when handled by farm workers or disturbed by cattle. Hay is used mostly during the winter to feed cattle indoors, where concentrations of spores in the air of cowsheds can be very large and farm workers are repeatedly exposed to them both daily and in successive years.

Dry hay does not heat and generate these large numbers of spores. Showery weather during hay-making can prevent hay drying adequately in the field, and farmers lacking labour and time, may then store the hay damp. The introduction of the pick-up baler has speeded hay making and considerably decreased the labour required. It also enables hay to be stored wetter in bales than was considered safe with stacks of loose hay where the risk of spontaneous ignition was greater. Such wet bales still heat but, in contrast to the interior of stacks of loose hay, aeration is adequate for thermophilic actinomycetes to grow rapidly.

For Farmer's Lung to develop workers must be exposed often to many spores. These conditions are also fulfilled in other agricultural environments and are often associated with similar hypersensitivity diseases. Mushroom compost contains many thermophilic actinomycetes, and enough spores are released by workers handling the compost in closed sheds to cause the respiratory complaint known as Mushroom Worker's Lung (Bringhurst, Byrne and Gershon-Cohen, 1959; Sakula, 1967; Jackson and Welch, 1970). Large concentrations of spores of fungi, actinomycetes and bacteria, can also develop in grain silos that are not adequately sealed, and when the grain is removed at intervals spore clouds are generated around the delivery chute (Lacey, 1971*b*).

The thermophilic actinomycetes responsible for Farmer's Lung are common in soil and will contaminate any agricultural product. When the products overheat during storage these species proliferate and produce many spores. Future changes in agricultural practice, and alternative methods of storing fodder, may ease the task of the farmer but may also lead to conditions even more conducive for microorganisms to grow rapidly and profusely.

These thermophiles also infect sugar cane, and changes in the industry are producing increased hazards. Until recently much of the residue after crushing the cane and extracting the sucrose (bagasse) was burnt, but the fibrous bagasse is now in demand for various purposes. Spores from soil can germinate in the stored bagasse when it is wet enough, generate heat and produce as large concentrations of spores as in hay bales. These are released when the bales are broken open and can cause the disease bagassosis (Jamison and Hopkins, 1941; Buechner *et al.*, 1958; Salvaggio *et al.*, 1966; Hearn, 1968; Lacey, 1971*a*).

Harvesting of hay and grain can also generate large concentrations of dust in the field, especially when the season has been wet. Respiratory complaints have been reported in farm workers engaged in harvesting, and people have been found who are allergic to grain dust and to the phytopathogenic smut fungi (*Ustilago* spp) that are often present (Harris, 1939).

Industry

Sugar cane bagasse provides a link between agricultural and industrial situations, for bagassosis is characteristically an industrial disease associated with the use of bagasse to make hardboard, paper and insulating materials. Other examples occur where agricultural products are handled industrially, so that some workers become sensitized to various dusts associated with seeds, textile fibres, woods and gums. Such workers can develop acute asthma even when the dust in the atmosphere is very dilute. The attacks characteristically stop when the worker is no longer exposed to the dust, and recur when he is again exposed.

Respiratory allergy may occur in workers handling grain. Allergy to grain dust, and to some of the component fungi has been found in dockers, elevator operators and mill workers (Dunner, Hermon and Bagnall, 1946; Jiminez-Diaz, Lahoz and Cento, 1947; Ordman, 1958; Skoulas, Williams and Merriman, 1964; Williams, Skoulas and Merriman, 1964). Heavy contamination of grain in maltings by *Aspergillus fumigatus* or *Aspergillus clavatus* can cause a disease resembling Farmer's Lung in workers moving the grain (Vallery-Radot and Giroud, 1928; Filip and Barborik, 1966; Riddle *et al.*, 1968).

Cryptostroma corticale is a phytopathogenic fungus that grows under the bark of maple trees. When the bark is removed in sawmills, spore clouds may be generated that can cause asthma and allergic alveolitis in workers (Towey, Sweaney and Huron, 1932; Emanuel, Lawton and Wenzel, 1962; Wenzel and Emanuel, 1967). Other fungi that mould cork and redwood sawdust cause the diseases suberosis and sequoiosis, respectively, in workers handling them (Ávila and Villar, 1968; Cohen *et al.*, 1967).

Fungus spores have been implicated in a respiratory complaint resembling Farmer's Lung in factory workers engaged in producing citric acid (Hořejši *et al.*, 1960). Trays of nutrient medium with surface growths of *Aspergillus niger* often become contaminated by *Aspergillus fumigatus* and various *Penicillium* spp. Dense clouds of spores are released during harvest at intervals of 3 days, and workers were not protected by wearing cotton wool filters. Spores were isolated from sputum and laryngeal swabs, skin sensitivity tests were positive and circulating antibodies against the fungi being handled were found. Complaints by the workers were most common when the cultures were contaminated by "green spores", and the contaminating *Penicillium* spp were found to be important in the development of symptoms.

The pattern is very similar to Farmer's Lung in agricultural workers. Intermittent but frequent exposure to large spore concentrations in poorly ventilated areas produced hypersensitivity symptoms in a proportion of the workers. However, air-conditioning systems may also prove a mixed blessing. A recent account of aspergillosis in a laboratory working with synthetic fibres described large concentrations of spores in the air caused by massive growths of *Aspergillus fumigatus* in the air-conditioning system (Wolf, 1969). Respiratory symptoms resembling Farmer's Lung have also been caused by the actinomycetes, *Micropolyspora faeni* and *Thermoactinomyces vulgaris*, growing in humidifiers in air-conditioning units (Banaszak, Thiede and Fink, 1970; Luedemann, pers. comm.). Spores may be rapidly spread through a building by air-conditioners, especially where the organism grows within the system, and may affect people remote from the source of spores.

A classical but baffling example of such a condition is byssinosis. This respiratory complaint is evident in cotton-mill workers returning to work after the weekend holiday, when they suffer from "flu-like" symptoms, accompanied by fever, asthmatic dyspnoea, constriction of the chest and general malaise (Schilling *et al.*, 1955). Ordinarily the onset is acute and the illness brief, but continued exposure and repeated attacks can lead to impairment of pulmonary function. Byssinosis is particularly associated with exposure to the dust of cotton, flax and hemp, and does not develop in workers handling sisal, jute and man-made fibres. The ingredients of the dust causing byssinosis have been variously described as the plant protein associated with the fibres, histamine, histamine liberators, endotoxins, fungal spores or "various infective agents that can give rise to bronchiolar constriction". The reactions may possibly be precipitin mediated (Pepys, 1969). Crofton and Douglas (1969) consider Weaver's Cough to be a disease similar to Farmer's Lung caused by mouldy cotton.

One can predict that large spore concentrations associated with novel industrial processes may cause similar problems in the future. Fungal spore suspensions for the transformation of steroids and other pharmacological compounds have been suggested and used in pilot scale experiments. The generation of aerosols or dry spore clouds during the preparation of such spore suspensions must be prevented. The cultivation of fungi on moist carbohydrate foods to increase their protein content and palatability has been advocated as a means of providing alternative food supplies. Carbohydrates of cereals, manioc, yam and sweet potato would be fermented with chosen fungal strains, so improving on the traditional fermented foods such as *tempeh*, *sufu* and *miso*. Another possible future development is the use of fungi to prepare entirely new products. Combinations of cereals and legumes fermented to give a product with a desirable nutritional balance and taste might prove more acceptable than cakes of single cell protein. Such large scale future developments in the production of fermented foods could result in the generation of large spore concentrations in buildings with restricted ventilation.

Microbial products may also be antigenic and cause symptoms of allergy when inhaled. An epidemic outbreak of systemic "influenza" in 14 workers engaged in the production of tuberculin was shown to be due to the inhalation of a tuberculin aerosol (Radonic, 1966). Respiratory complaints have been reported in workers inhaling dust containing commercial enzyme preparations of microbial origin (Flindt, 1969, Pepys *et al.*, 1969), and also in some users of washing powders containing these enzymes (Belin *et al.*, 1970). From these examples microbial products should be suspected of causing allergy as well as the spores.

Laboratory

Most authors of articles on laboratory safety have concentrated on the dangers of aerosols carrying pathogenic bacteria (e.g. Wedum and Kruse, 1969). Darlow (1969)—see pp. 1–20—listed some basic rules to guide laboratory workers that include: "always regard all microorganisms as potentially pathogenic" and "ensure that all cultures are sterilized before they are washed up". It is surprising how often "all pathogens" is translated as a few pathogens. Fungal cultures, with the possible exception of dermatophytes, are often handled without any regard for their possible harmful effects. Workers may also realize the dangers associated with spore clouds of *Aspergillus fumigatus* and handle pure cultures with care, but isolation plates carrying large grey-green sporulating fungus colonies can litter laboratory benches for long periods. We have seen the agar from such glass Petri dishes discarded into a bucket for later disposal, so producing clouds of spores. When this is done frequently in the enclosed space of a washing-up room, large concentrations of airborne spores from various species may be generated frequently.

Repeated exposure to spores of particular species may occur in laboratories engaged in genetical research, or in the testing of mould resistance of materials or equipment or the effects of biocides using standard test systems. Often the risk may be greater from contaminants, such as *Aspergillus fumigatus*, than from the test organisms.

Workers in control laboratories receiving samples of agricultural or manufactured materials affected by microbial spoilage may liberate spores when the containers are opened and the contents processed. We know of no specific example of laboratory workers suffering from spore-induced hypersensitivity, but Harington (1967) also suggested that the dangers do exist. He specifically proposed that workers examining moulds, foodstuffs, and grain for the presence of mycotoxins were at risk. North and Gwynne (1960) concluded there were no toxic hazards from large scale culturing of the fungus, *Pithomyces chartarum*, cause of facial eczema in sheep. However, despite reporting large concentrations of airborne spores and symptoms suggestive of allergy, they neglected this possibility.

Allergic Respiratory Disease caused by Inhaled Organic Dusts

Allergy was defined by von Pirquet (1906) as the "**acquired, specific, altered capacity to react**".

Sensitivity is **acquired** by adequate exposure to allergens, leading to the production of antibodies. Avoiding or limiting exposure prevents or limits sensitization. The antibodies are evidence of exposure to the particular allergen and, until their clinical relevance has been determined, cannot be regarded as diagnostic.

Reactions between antigen and antibody are **specific**, depending on immunochemical specificity. However, "false positive" reactions may occur. For example, clinical "false positive" reactions may occur to antigens unrelated to the source of the specific dust or to the disease, such as to teichoic acids from *Staphylococcus aureus*, a common contaminant of organic dusts, against which a large proportion of the population has antibodies. Non-immunological "false positive" reactions may also occur, such as the precipitation of C-substance glycopeptides in bacteria, fungi, parasites and vegetable dusts by C-reactive protein in the serum. This reaction can be prevented by using a calcium chelating agent such as sodium citrate.

Sensitization is shown by an **altered capacity to react**, that is by the production of allergic symptoms on exposure to substances previously without such effects, and by preventing symptoms by avoiding exposure.

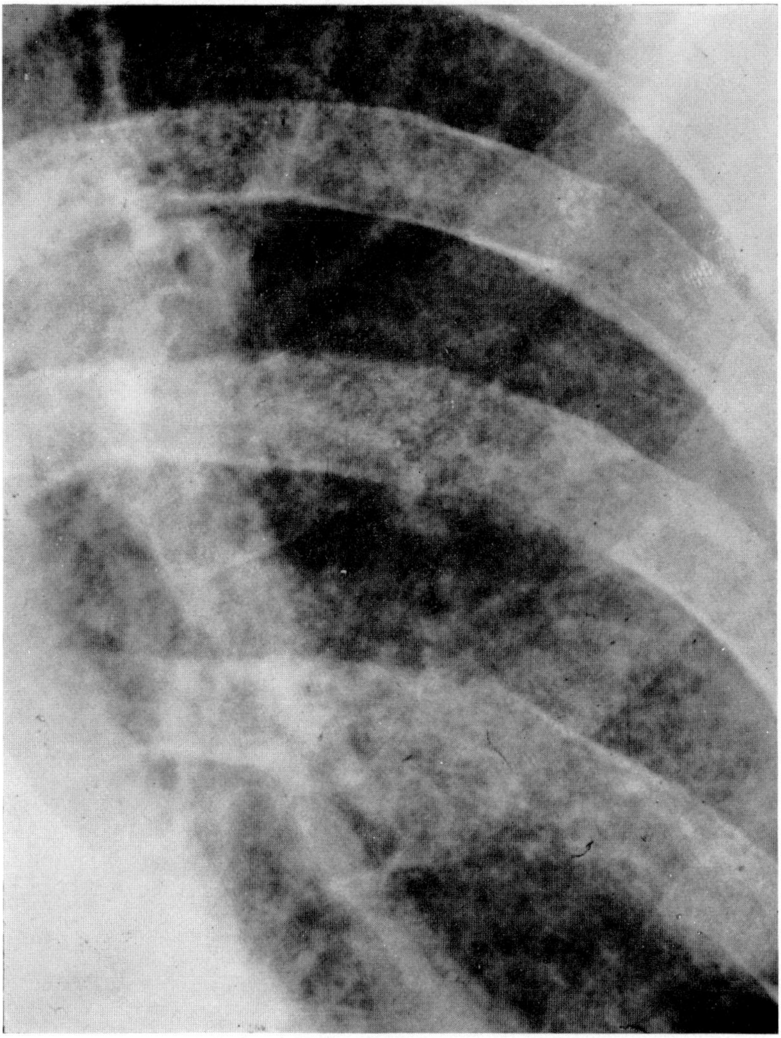

FIG. 1. Radiographic appearance in extrinsic allergic alveolitis (Farmer's Lung). Whitish mottled shadowing shows widespread micronodular infiltrations in a patient with Farmer's Lung caused by inhalation of *Micropolyspora faeni*, against which precipitins were present.

TABLE 1. Comparison of the features of allergic sensitivity in atopic and non-atopic subjects.

Type of allergy Mainly	"Atopic" subjects Immediate, Type I	"Non-atopic" subjects Arthus, Type III
Sensitization	Constitutional predisposition to be readily sensitized by ordinary exposure to daily life.	More extensive exposure required.
Antibody	Reaginic antibody, IgE.	Precipitating antibody IgG and IgM.
Mechanisms of reaction	Combination of allergen with reaginic antibody attached to surface of mast cells, causes rapid release of histamine and other mediators of tissue reactions.	Immune complexes, of allergen and antibody formed, which fix and activate components of complement enzymatically.
Tissue response	Oedema, eosinophilia, hypersecretion, reversible and not tissue damaging.	Perivascular inflammatory response, mononuclear cells mainly, granuloma formation, fibrosis, tissue-damaging reaction.
Speed and duration of reaction typically	Rapid, "immediate" onset in minutes, lasts $1\frac{1}{2}$ to 2 h, as seen on skin and inhalation tests (Figs 12 and 14).	Slow, "late" onset in about 3 to 4 h, lasts 24 to 36 h, as seen in skin and inhalation tests (Figs 13 and 15).
Upper respiratory tract and eyes	Rhinitis, conjunctivitis, itching, sneezing, rhinorrhoea, nasal obstruction, lacrymation.	Not a well-defined feature.

Bronchi	Asthma—rapid onset. Immediate, Type I, reaction. Wheezing and dyspnoea (Fig. 14).	(a) Asthma—slow "late" onset, Type III reaction. Wheezing and dyspnoea.
Peripheral gas-exchanging, "alveolar" tissues	Not affected	(b) Extrinsic allergic alveolitis (Farmer's Lung type of disease), Type III reaction. Dyspnoea, cough, no wheezing, crepitant rales present. (a) and (b) may occur together in some subjects.
Systemic reactions	Eosinophilia of blood tissues and secretions. Nothing else characteristic.	Eosinophilia not a feature. Fever, malaise, myalgia and loss of weight which may be considerable.
Pulmonary function	Ventilatory obstruction—expiration decreased and prolonged.	Restrictive ventilatory effect—decrease in gas-transfer and elasticity of the lungs.
Radiography of lungs	Nil characteristic.	Extrinsic allergic alveolitis—miliary nodular infiltration, fibrosis and cystic changes (Fig. 1).
Pathology	Oedema and eosinophilia of bronchial mucosa.	Infiltration of alveolar walls with lymphoid histiocytic and plasma cells. Epithelioid cell granulomata, fibrosis and cystic changes (Fig. 2).

Because the type of antibody determines the form of allergic reaction, different allergens can produce the same type of allergic reaction and the same clinical manifestations.

The production and nature of allergic disease is determined by three factors: the immunological reactivity of the subject, the nature of the inhaled particles, and the circumstances of exposure.

Immunological reactivity of the subject

The population divide into two groups, known respectively as atopic and non-atopic, depending on their characteristic pattern of allergic response. The atopic group is composed of *c.* 10% of the population who have a constitutional predisposition to become easily sensitized by ordinary exposure of everyday life. Table 1 summarizes the main features of the atopic and non-atopic groups of subjects with their different forms of sensitivity. Although Type I allergy is, in this context, characteristic of the atopic group and Type III allergy of the non-atopic group, the different types of allergy can occur together and may even be interdependent.

Aspergillus fumigatus provides examples of the effects on allergic manifestations and infection of the two patterns of sensitivity; namely, Type I allergy mediated by reaginic IgE antibody in the atopic subject, and Type III allergy mediated by precipitins in the non-atopic subject.

Atopic subjects may develop

Asthma. Develops showing immediate, Type I skin and asthmatic reactions to spores and extracts of *Aspergillus fumigatus.*

Allergic broncho-pulmonary aspergillosis. The Type I reaction described above may be complicated by the appearance of precipitins and Type III allergy, resulting in one of the forms of "pulmonary eosinophilia" together with the asthma. Transient recurrent pulmonary infiltrations appear (Fig. 3) together with the expectoration of tough "plugs" of sputum containing mycelium of *Aspergillus fumigatus* and eosinophil cells (Fig. 4). Precipitin tests are usually positive, but weak, and may require concentration of the serum (Fig. 5). Inhalation tests give a dual reaction; that is, an immediate asthmatic response mediated by IgE antibody, followed later by a "late", Type III asthmatic reaction with fever and leucocytosis, mediated by precipitating antibody.

Non-atopic subjects may develop

Extrinsic allergic alveolitis. Large concentrations of spores of *Aspergillus fumigatus*, and of the other *Aspergillus* spp, may cause reactions of the peripheral gas-exchanging tissues of the lungs in non-atopic subjects who

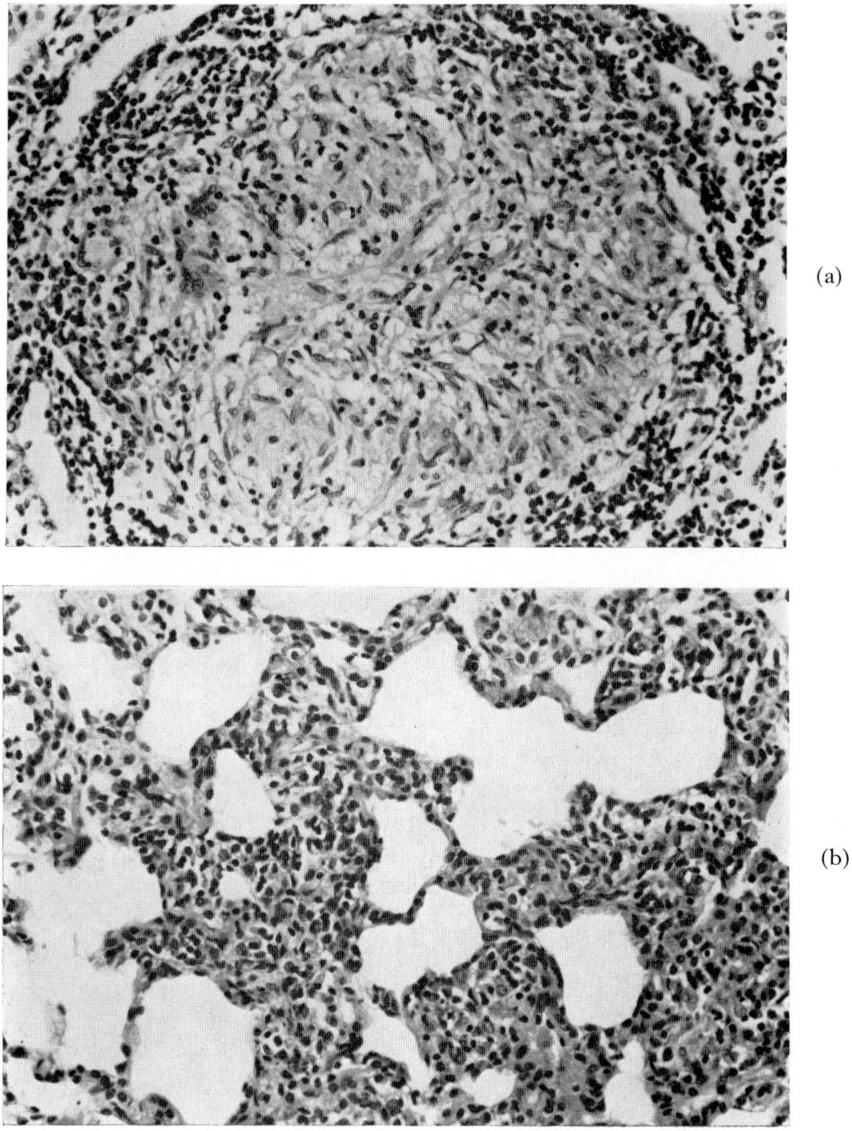

FIG. 2. Pathology of extrinsic allergic alveolitis. (a) Epitheliod cell granuloma and giant cell surrounded by lymphoid cells in the alveolar wall in Farmer's Lung, and (b) infiltration of alveolar walls with lymphoid plasma and histiocytic cells in Bird Fancier's Lung.

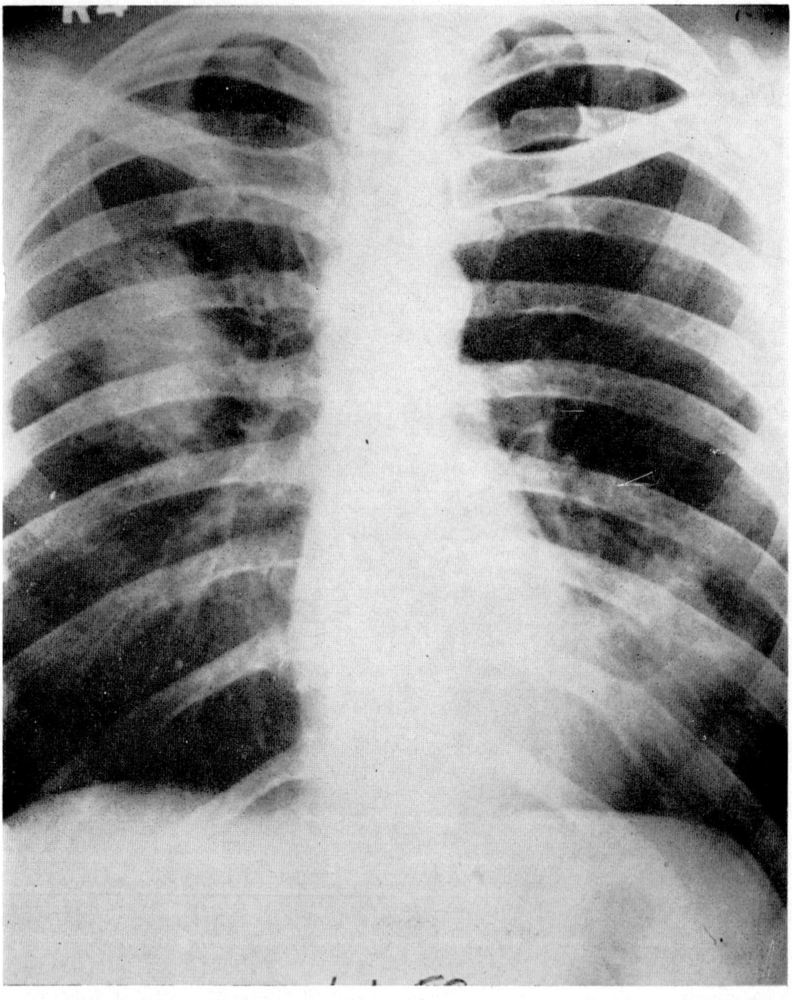

FIG. 3. Radiographic appearance in allergic bronchopulmonary aspergillosis, showing peribronchial infiltrations.

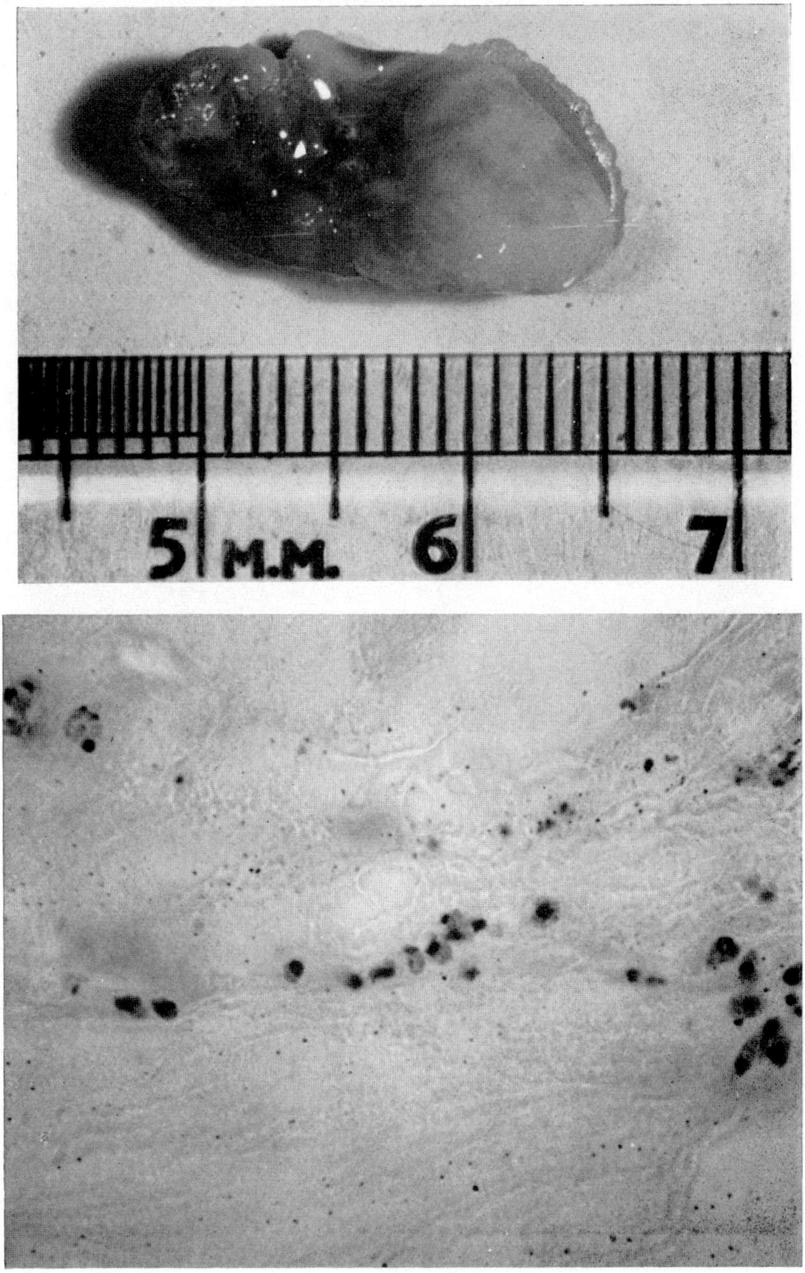

(a)

(b)

FIG. 4. (a) Sputum plug from a patient with allergic bronchopulmonary aspergillosis; (b) Eosinophil cells in the plug.

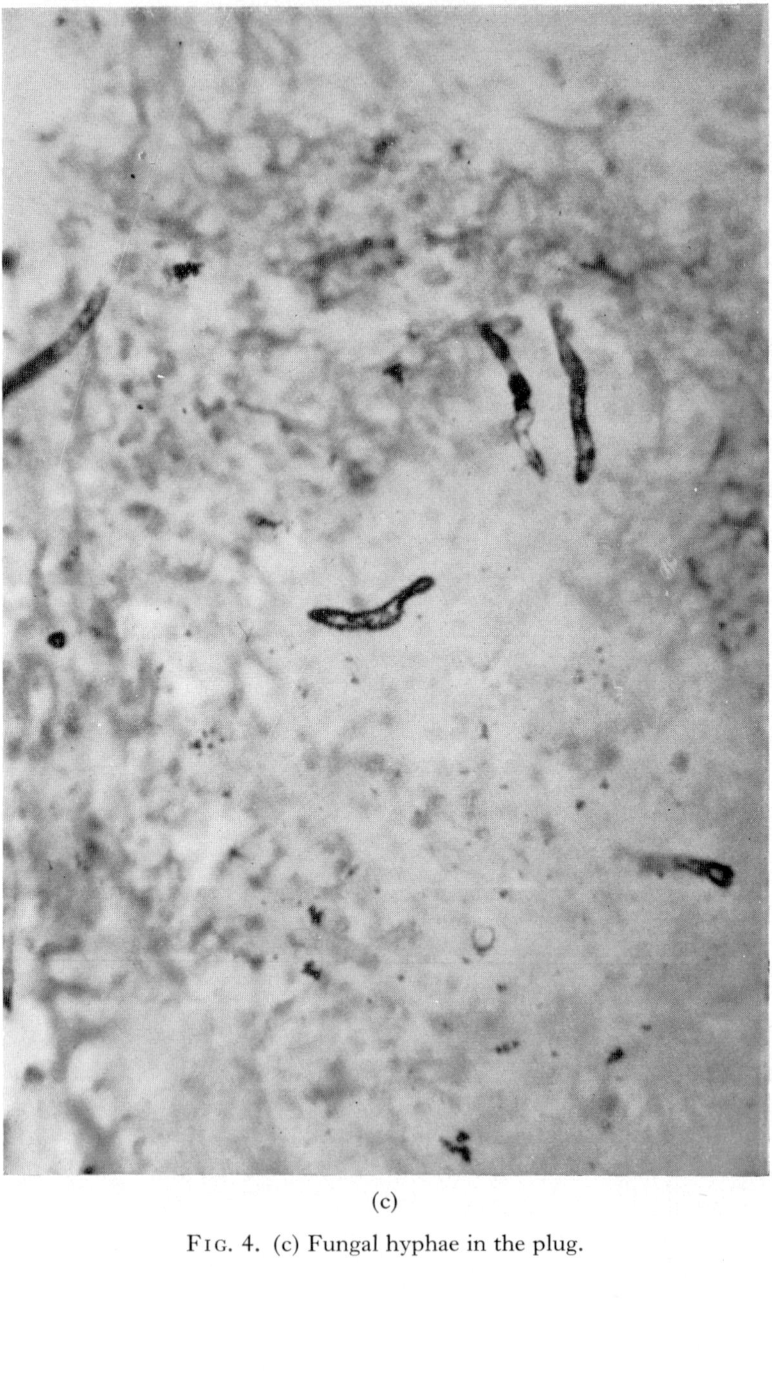

(c)

FIG. 4. (c) Fungal hyphae in the plug.

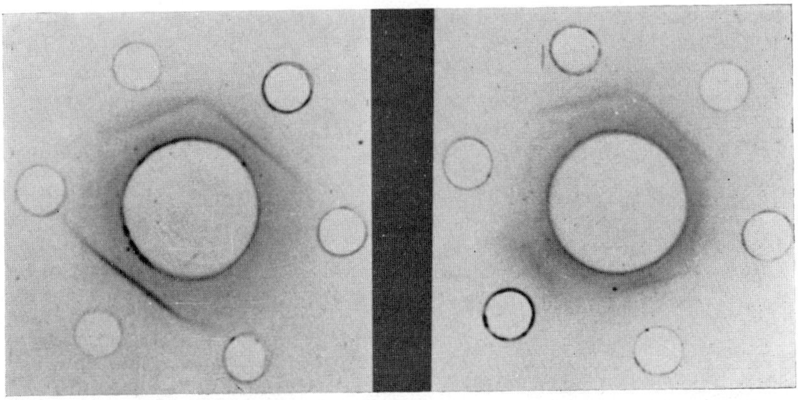

FIG. 5. Agar-gel double diffusion precipitin test in allergic bronchopulmonary aspergillosis. Typical positive, but weak reactions to some extracts of *Aspergillus fumigatus* in the peripheral wells, tested against the sera of two patients in the centre wells. It is desirable to use more than one extract to obtain maximum positive reactions, because of differences in antiboides in different patients and of antigens in different extracts.

have developed precipitins, with the development of extrinsic allergic alveolitis.

Aspergilloma. Inhaled spores of *Aspergillus fumigatus* may germinate and grow saprophytically in areas of damaged lung, resulting in the fungus ball or aspergillus mycetoma (Fig. 6). Precipitin reactions are typically strongly positive with the formation of multiple arcs (Fig. 7).

Invasive aspergillosis. Aspergillus fumigatus may invade the tissues, in reticulo-endothelial disease, during immunosuppressive drug therapy, or after massive X-ray irradiation, and precipitin tests may be positive.

An example of the different effects of *A. fumigatus* spores in atopic and non-atopic subjects is provided by two laboratory workers who were accidentally heavily exposed when culture plates were broken. One, a non-atopic subject, felt mildly ill for 2 to 3 days and had a slight cough for a further 10 days; skins and serological tests were negative. The other, an atopic subject, developed allergic aspergillosis of the sinuses and antra, and later of the bronchi; skin tests were positive and precipitins were present.

The nature of inhaled particles

The response to inhaled particles is affected by their size, chemical composition and other characteristics—see also p. 6. Spores of different species of bacteria, actinomycetes and fungi range from shorter than 1 μm

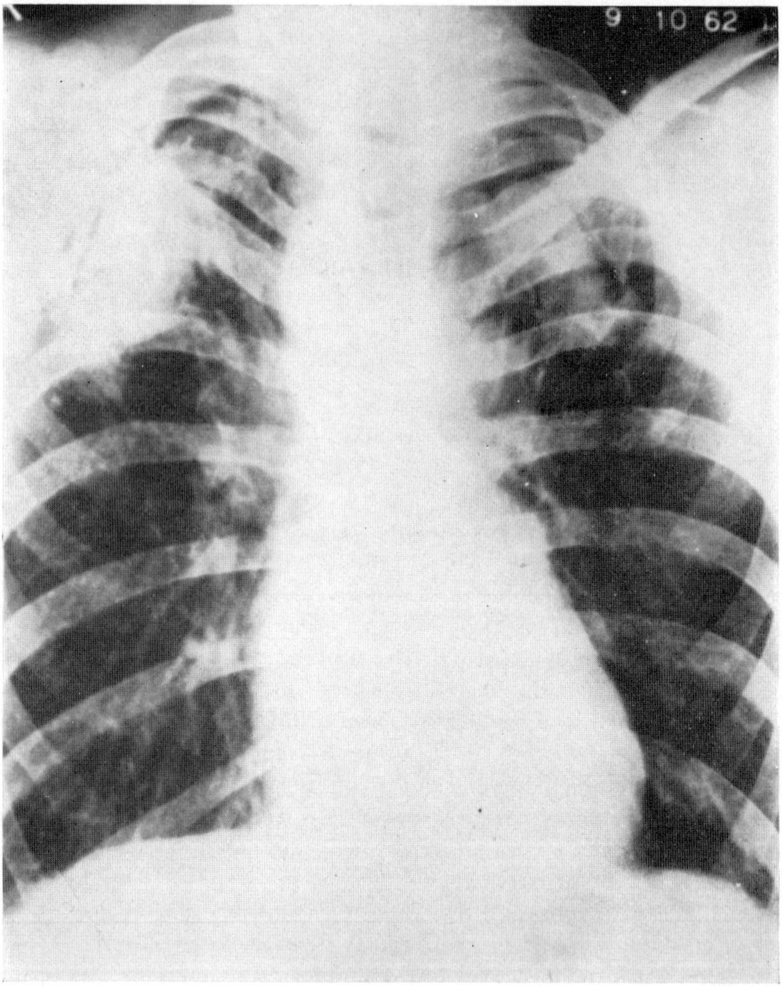

FIG. 6. Radiographic appearance of aspergilloma. Two "fungus balls", one in the right and the other in the left upper zones, surrounded by typical halo of air. This is a saprophytic growth in pre-existing lung cavities, commonly from healed pulmonary tuberculosis, and occurs mainly in non-atopic subjects.

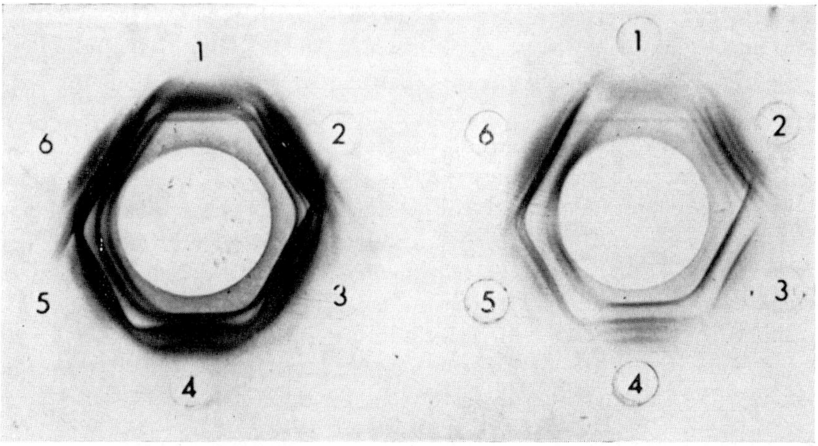

FIG. 7. Agar-gel double diffusion precipitin test in aspergilloma. Typical strong reaction with multiple arcs to different extracts of *Aspergillus fumigatus* in the peripheral wells. Differences can be seen in antigens of different extracts of the same strain prepared from different cultures.

to larger than 100 μm. However, spores of a given species vary only within a narrow size range, in contrast to many inorganic dusts. The size of a spore determines how far into the respiratory tract it will penetrate before it is deposited and the type of allergy it may provoke. Most spores of a given species will thus be deposited in the same region of the respiratory tract and provoke similar symptoms. Figure 8 shows the approximate relationship between spore size, site of deposition and allergic symptom

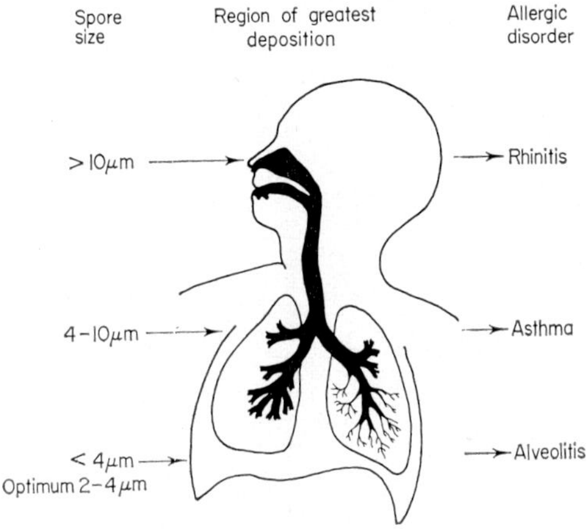

FIG. 8. Spore, size, lung penetration and type of allergic disease.

caused. This cannot be regarded as precise, as there is no sharp cut off from one region to another, and some spores, such as *Aspergillus fumigatus*, tend to aggregate so their effective particle size is greater than that of a single spore.

During normal nose breathing, particles larger than 10 μm diam. are trapped in the nose, where they may cause rhinitis, and those smaller than 5 μm mostly penetrate to the alveoli, where they may cause alveolitis. Those intermediate are mostly deposited in the bronchi and bronchioles, and cause asthma, but when concentrated enough, may penetrate to the alveoli to cause alveolitis. The optimum size for alveolar deposition is 2 to 4 μm (Muir, pers. comm.). Breathing through the mouth enables particles larger than 10 μm to enter the lungs, be deposited in the trachea and bronchi, and cause asthma instead of rhinitis (Lidwell, 1970).

TABLE 2. Fungi and actinomycete species, grouped according to spore size, showing the type of disease they may cause

Spore type	Disease: Rhinitis	Asthma	Alveolitis	Infection
Spores larger than 10 μm				
Alternaria tenuis	+	+		
Epicoccum nigrum	+			
Puccinia graminis	+	+		
Grass pollen	+	+		
Spores 4–10 μm				
Alternaria tenuis	+	+		
Arthrinium phaeospermum	+	+		
Aspergillus flavus	+	+		+
Aureobasidium pullulans			+	
Cladosporium fulvum	+	+		
Cladosproium herbarum	+	+		
Graphium sp			+	
Mucor pusillus	+			+
Penicillium roqueforti	(+)	(+)	(+)	
Serpula lacrymans	+	+		
Sporobolomyces roseus	+	+		
Ustilago avenae	+	+		
Ustilago hordei	+	+		
Ustilago nuda	+	+		

Spore type	Disease Rhinitis	Asthma	Alveolitis	Infection
Spores 2–4 μm				
Absidia corymbifera				+
Absidia ramosa				+
Arthrinium phaeospermum	+	+		
Aspergillus clavatus	+	+	+	
Aspergillus flavus	+	+		+
Aspergillus fumigatus	+	+	+	+
Aspergillus niger	+	+	+	+
Aspergillus terreus	+	+		+
Aureobasidium pullulans			+	
Cryptostroma corticale			+	
Graphium sp			+	
Mucor pusillus	+	+		+
Penicillium casei	+	+	+	
Penicillium caseicolum	+	+	+	
Penicillium herquei	+	+	+	
Penicillium miczynski	+	+	+	
Penicillium piceum	(+)	(+)	(+)	
Penicillium roqueforti	(+)	(+)	(+)	
Penicillium rubrum	+	+	+	
Penicillium simplicissimum	+	+	+	
Sporobolomyces roseus	+	+		
Spores less than 2 μm				
Aspergillus terreus	+	+		+
Micropolyspora faeni		+	+	
Nocardia asteroides				+
Thermoactinomyces vulgaris		+	+	
Thermoactinomyces sacchari			+	

(+), Potential but unproven cause of disease. Note, when a species has been identified as the cause of a particular disease, the other species of that genus should be regarded as potential causes, even though not yet proved.

Table 2 shows examples of fungi and actinomycetes commonly encountered during harvesting, in mouldy fodders, bagasse, mixtures of spores for mould resistance tests or in outdoor air, grouped according to size and showing the type of disease they may cause. Not all have yet been proved to cause a disease, but where one species of a genus has been shown to cause an allergy, the others are shown likewise. All spores, whether viable or not, may be regarded as potential allergens. Figures 9 and 10 illustrate representative spores, and Table 3 lists species causing recognized forms of extrinsic allergic alveolitis.

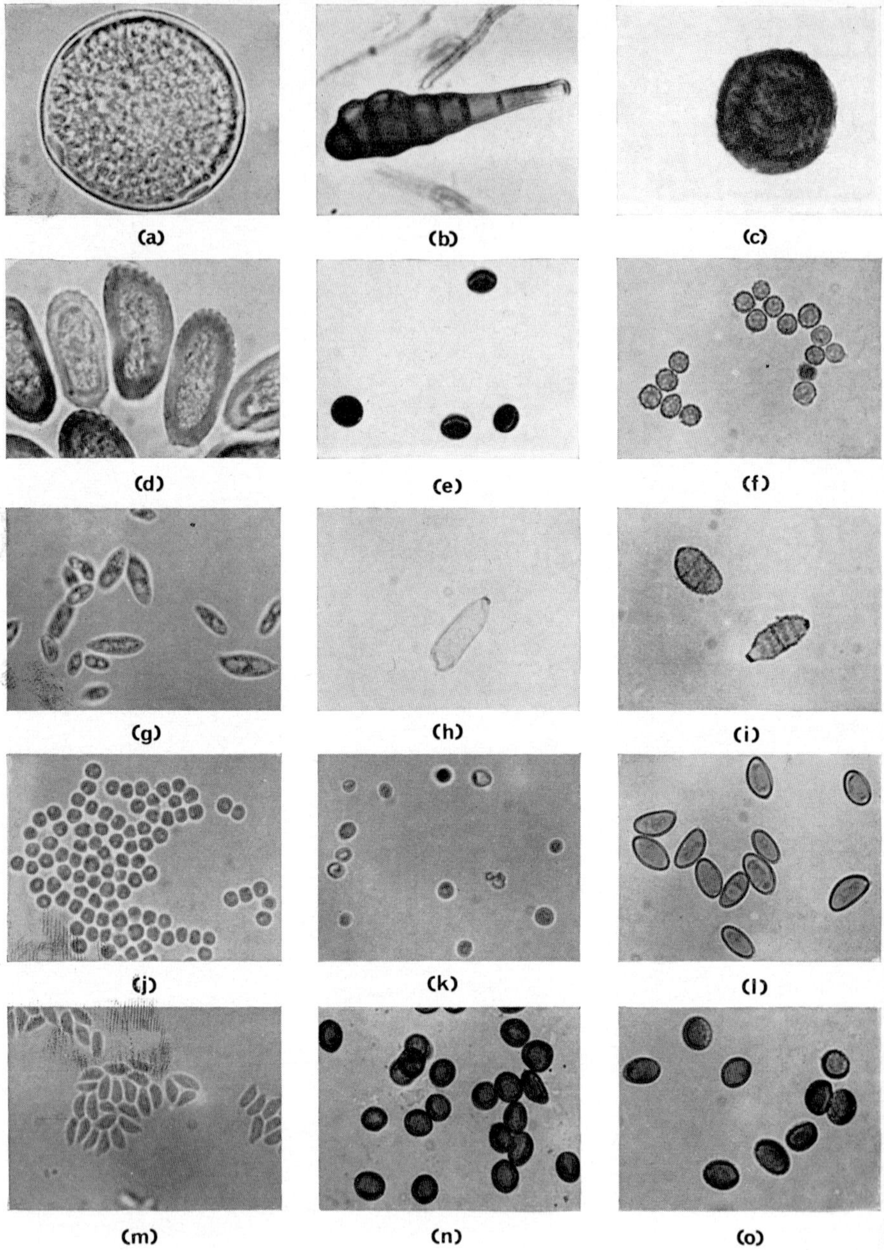

FIG. 9. Examples of fungus spores in different size ranges compared to grass pollen (a) *Phleum pratense* greater than 10μm, (b) *Alternaria tenuis*, (c) *Epicoccum nigrum*, (d) *Puccinia graminis*. From 4 to 10μm: (e) *Arthrinium phaeospermum*, (f) *Aspergillus flavus*, (g) *Aureobasidium pullulans*, (h) *Cladosporium fulvum*, (i) *Cladosporium herbarum*, (j) *Penicillium roqueforti*, (k) *Mucor pusillus*, (1) *Serpula lacrymans*, (m) *Sporobolomyces roseus*, (n) *Ustilago avenae*, and (o) *Ustilago nuda* (all × 650).

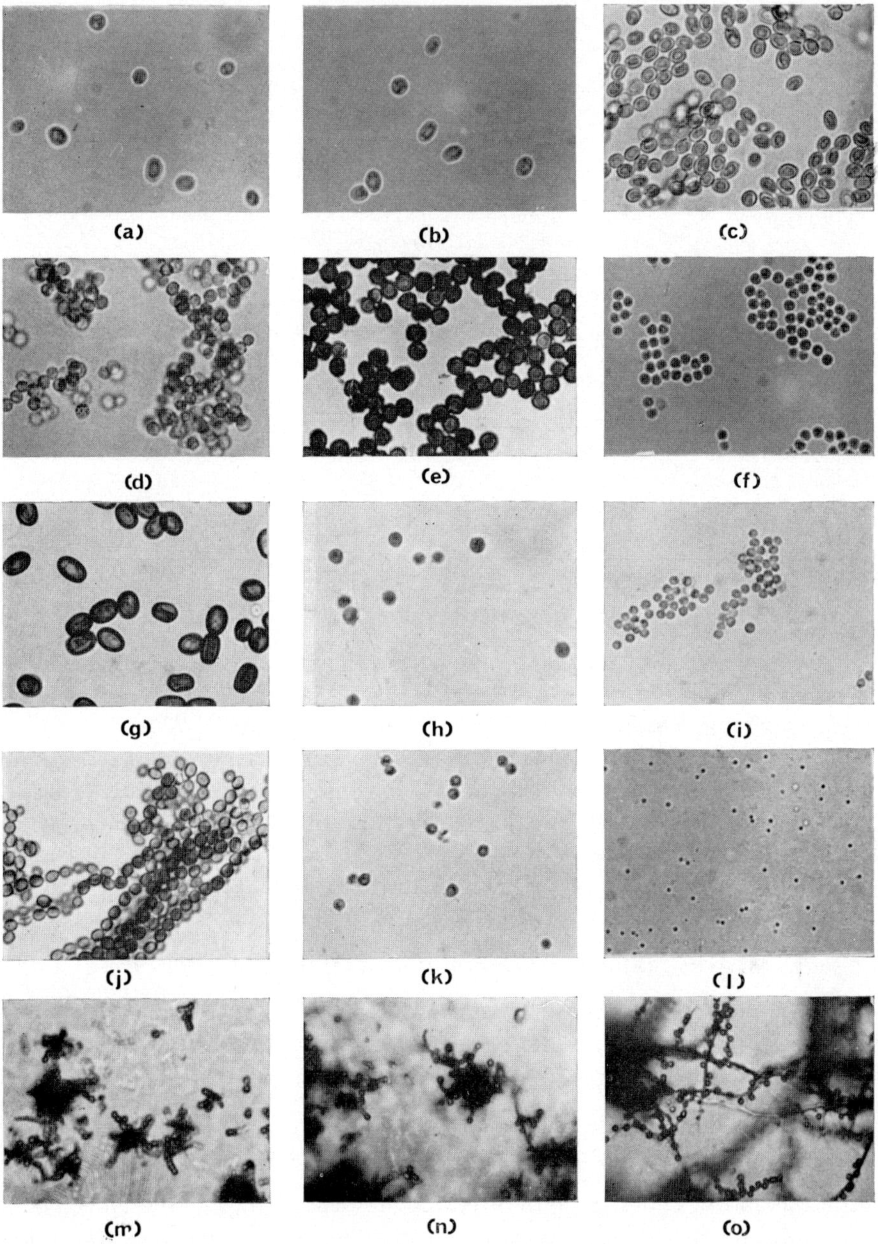

FIG. 10. Examples of spores in different size ranges (continued). From 2 to 4μm: (a) *Absidia corymbifera*, (b) *Absidia ramosa*, (c) *Aspergillus clavatus*, (d) *Aspergillus fumigatus*, (e) *Aspergillus niger*, (f) *Aspergillus terreus*, (g) *Cryptostroma corticale*, (h) *Penicillium casei*, (i) *Penicillium miczynskii*, (j) *Penicillium piceum*, (k) *Penicillium simplicissimum*. Smaller than 2μm: (l – o) actinomycete spores, (l) slide preparation, (m – o) growing on aerial mycelium, (m) *Micropolyspora faeni*, (n) *Thermoactinomyces sacchari*, and (o) *Thermoactinomyces vulgaris* (a–l, × 650; m–o, × 500)

TABLE 3. Names applied to extrinsic allergic alveolitis caused by inhaled spores

Source of dust	Organism	Disease
Mouldy hay	*Micropolyspora faeni* *Thermoactinomyces vulgaris*	Farmer's Lung
Air conditioning systems	*Micropolyspora faeni* *Thermoactinomyces vulgaris*	Hypersensitivity pneumonitis
Bagasse	*Thermoactinomyces sacchari*	Bagassosis
Redwood sawdust	*Aureobasidium pullulans* *Graphium* sp	Sequoiosis
Malting barley	*Aspergillus clavatus* *Aspergillus fumigatus*	Maltworker's Lung
Maple bark	*Cryptostroma corticale*	Maple bark pneumonitis
Cheese	*Penicillium casei*	Cheese washer's Lung

Circumstances of exposure

The way in which a subject is exposed to an allergen will affect sensitization and the way in which the disease develops. The greater the intensity of exposure, the faster and the greater the sensitization. With extrinsic allergic alveolitis, intermittent exposure causes attacks with an acute onset 4 to 5 h after exposure, so that the relationship of exposure to the disease is evident. By contrast, regular exposure even to small concentrations allows the disease to develop insidiously and more dangerously, because the relationship between exposure and the disease is much less obvious.

Investigating an Outbreak of Respiratory Allergy

When fungus or actinomycete spores are suspected of causing a disease the source of antigens can best be identified by parallel investigations of the air spora and the immunology of exposed workers.

Determination of the air spora

No method of spore trapping gives a complete picture of the air spora. Methods relying on a microscopic assessment of the catch enable the total spore concentrations to be estimated, but relatively few types can be identified even to genus (Lacey, 1971*b*). Methods based on growing trapped spores on agar media allow only viable spores to be estimated, and may be affected by inter-colony competition, selectivity of the medium and incubation temperature. Species growing weakly under a given set of cultural conditions may be overlooked, but may nevertheless be important in the etiology of the disease.

A combination of the cascade impactor (May, 1945), allowing microscopic assessment, and the Andersen sampler (Andersen, 1958), allowing growth in culture—for additional details, see p. 37—has proved invaluable in assessing the air spora of farm buildings (Fig. 11) (Lacey and Lacey,

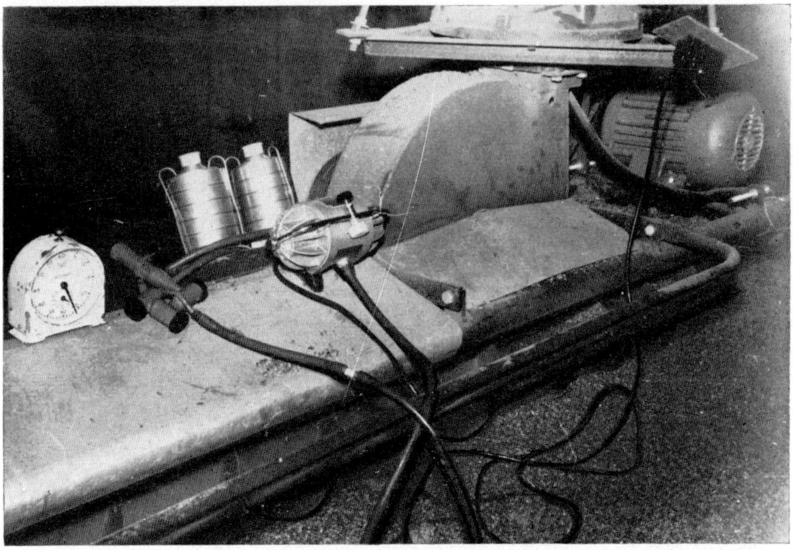

FIG. 11. The cascade impactor and Andersen sampler being used for air sampling on a farm.

1964; Lacey, 1969, 1971*b*). Both instruments separate the particles trapped into different size grades that can be related to the probability of penetration to different parts of the respiratory tract.

Other instruments that can be used to assess the air spora microscopically include the automatic volumetric spore trap (Hirst, 1952) and the "Casella" personal sampler; and to give growth in culture, the slit sampler (Bourdillon, Lidwell and Thomas, 1941) and liquid impingers of various types—for additional details, see p. 37— including a three stage form that separates particles according to size (May, 1966).

The usefulness of different instruments for growing microorganisms depends on the species involved and its source. The Andersen sampler enabled actinomycetes from hay, deposited dry on the surface of agar in Petri dishes, to be grown whereas they could not be grown from the trapping liquid in a liquid impinger, partly because bacterial growth was greater on plates prepared from the liquid impinger.

Two per cent (w/v) Malt Extract Agar containing 40 units streptomycin and 20 units penicillin per ml, and half-strength "Oxoid" Nutrient Agar containing 50 μg actidone per ml, are good general media for fungi and actinomycetes, respectively (Gregory and Lacey, 1963). When specific microorganisms are suspected of being in the air, more selective media can often be used. The temperature of incubation may be important in determining the range of species isolated. Thermophilic species may be missed unless incubation temperatures are warm enough. Incubation at about 40° for fungi, and 40° and 60° for actinomycetes, and about 25° for both, is necessary where there is any possibility of thermophiles being present. Tables 4 and 5 give typical results of trapping in some agricultural environments.

TABLE 4. Typical cascade impactor results from air sampling in agricultural environments (10^6 spores/m^3 air)

Spore type	Spore sample from:			
	Open shed while moving hay	Moist barley silo while unloading	Moist barley silo (undisturbed)	Mushroom farm while spawning
Actinomycetes and bacteria	1·28	1791·56	6·92	204·59
Total fungi	3·87	1070·67	1·00	0·11
Aspergillus glaucus/flavus	0·39	758·07	0·05	—
Other *Aspergillus*	2·15	67·80	0·14	0·04
Cladosporium	1·20	1·92	—	—
Mucoraceae	—	68·93	0·03	—
Humicola	—	2·21	0·23	0·07
Yeasts	—	153·92	0·08	—
Alternaria	0·07	1·41	—	—

When the nature of the air spora is known, typical colonies of the most abundant types should be isolated and multiplied pure to provide antigens for immunological tests.

Immunological investigations

Preparation of test extracts

Crude extracts are useful for preliminary tests and may even be more useful clinically than the purified allergenic components. They may be prepared from spores, mycelium or culture medium. The method of preparation depends on the material being used.

The usual practice is for spores and mycelium to be defatted with acetone overnight and then extracted for several days in a mildly alkaline saline

TABLE 5. Typical Andersen sampler results from air sampling in agricultural environments (Total colonies on 6 Petri dishes)

	Spore sample from: Open shed while moving hay	Moist barley silo while unloading	Moist barley silo (undisturbed)	Mushroom farm while spawning
60° Actinomycetes and bacteria				
Micropolyspora faeni	30	—	683	—
Thermoactinomyces vulgaris	32	20	30	—
Bacteria	15	30	—	13
40° Actinomycetes and bacteria				
Actinobifida chromogena (?)	—	—	—	897
Micropolyspora faeni	—	—	468	—
Streptomyces spp (grey)	3	—	—	30
Streptomyces spp (white)	55	—	280	72
Bacillus licheniformis	8	3	38	19
Other bacteria	48	Uncountable*	458	635
25° Actinomycete and bacteria				
Streptomyces spp (grey)	2	18	6	1
Streptomyces spp (white)	16	157	121	1
Streptomyces spp (yellow)	39	888	134	257
Bacteria	1696	616	970	100
40° Fungi				
Absidia spp	—	14	32	—
Aspergillus flavus	—	884	222	—
Aspergillus fumigatus	3	2400	15	123
Chaetomium thermophile	—	—	—	34
Humicola lanuginosa	1	—	12	—
Mucor pusillus	2	26	62	—
Thermoascus crustaceous	—	—	16	—
25° Fungi				
Absidia spp	31	14	7	—
Alternaria sp	8	—	—	—
Aspergillus flavus	—	2014	186	—
Aspergillus fumigatus	—	—	2	57
Aspergillus nidulans	1	—	—	1
Aureobasidium pullulans	9	—	5	—
Cladosporium spp	1200	—	—	—
Epicoccum nigrum	27	—	—	—
Penicillum spp	43	—	193	—
Yeasts	398	356	88	—

*Bacteria probably suppressing actinomycetes.

solution containing 0·5% phenol (Coca's fluid). The extract is then filtered, dialysed and preferably freeze-dried. Liquid culture media are also used after filtration, dialysis and freeze drying, without prior extraction in Coca's fluid. Cultures on agar media may provide potent allergens by repeated freezing and thawing to release metabolites of the culture that have diffused into the agar.

Purified antigenic components can be obtained by protein precipitation, thus separating "protein" and "polysaccharide" antigens, column fractionation and other methods.

Control extracts of the culture media are also necessary to detect reactions to their constituents that might cause misleading results.

Freeze drying is not essential, but provides a useful, though crude, quantitative basis. The freeze-dried extracts are reconstituted for testing at 1 to 30 mg/ml with sterile saline solution.

Methods of testing

Skin tests. These may be made on subjects not receiving antihistamine drugs, either by prick or intracutaneous methods. Prick tests are made by introducing the point of a fine needle superficially into the skin through a drop of reconstituted extract. About 1/3,000,000 ml of extract is introduced into the skin in this way. Providing the patient is not shown to be unduly sensitive, prick tests may be followed where necessary by intracutaneous tests, in which 0·01 to 0·02 ml of extract is injected into the skin.

Positive reactions to skin tests are of two types, corresponding to the nature of the allergy. Immediate, Type I, wealing reactions (Fig. 12) are obtained in atopic patients with reaginic antibody against the allergen. These appear within minutes and are maximal after 15 to 20 min. Type III "late" reactions (Fig. 13) are obtained in patients with precipitins. These may be elicited with prick tests, but larger doses and intracutaneous tests are usually required. The reaction starts after several hours and is usually maximal after 5 to 6 h.

Provocation tests. The simplest provocation test is by deliberate exposure to the suspect material as normally encountered, under carefully observed conditions. Nasal tests, where a drop of the extract is placed in the nostril, may also be useful. Where reaginic antibody is present itching, sneezing and rhinorrhoea develop within minutes i.e. there is an "immediate" nasal reaction. "Late" nasal reactions, coming on after some hours, may also be caused. Inhalation tests, using an aerosol of the test extract, may be used to demonstrate Type I and Type III allergy. The initial concentration of the aerosol must be carefully determined by finding the concentration that gives only a small or no immediate reaction in prick tests. Atopic subjects with Type I sensitivity give a rapid and "immediate" asthmatic reaction

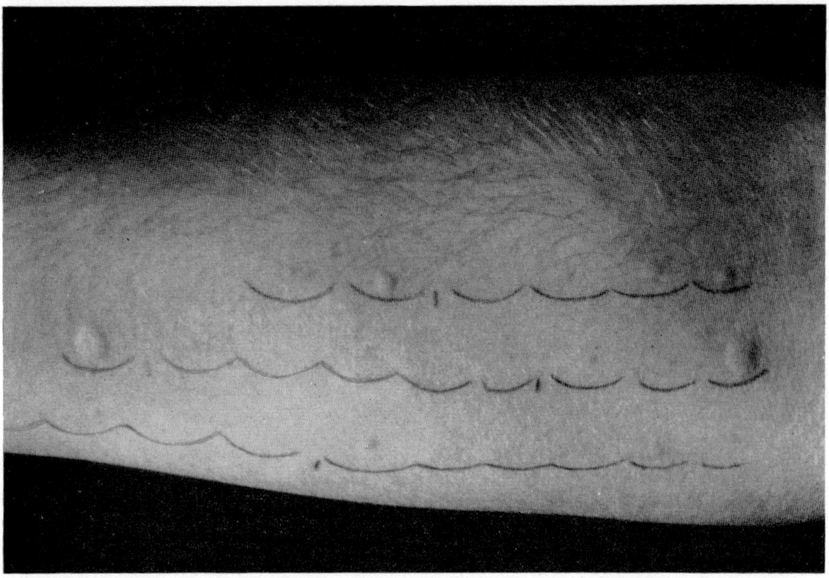

FIG. 12. Type I skin test reaction to common allergens. Multiple positive prick test reactions (weals after 15 to 20 min) in an atopic, asthmatic subject. Positive reactions: Inner aspect of forearm—No. 2, *Alternaria* +; No. 6, *Cladosporium herbarum* +; No. 1, Grass pollen ++++; 2, Flower pollens +; 3, Tree pollens +, and 9, *Dermatophagoides pteronyssinus* (House dust allergy mite) ++++. Outer aspect—No. 1, Cat dander ++; 2, Dog dander ++; 5, House dust ++, and 6, *Dermatophagoides farinae* +.

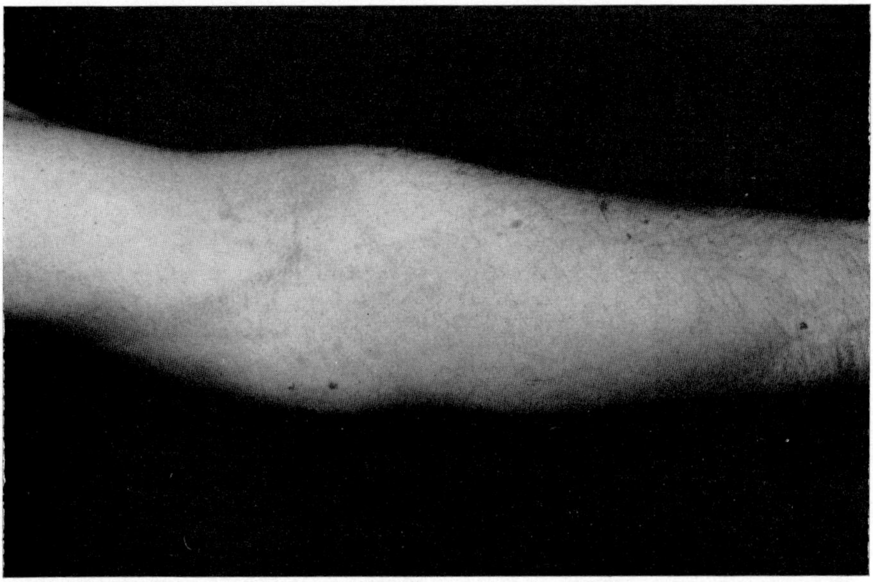

FIG. 13. Type III skin test reaction to *Aspergillus fumigatus* extract. Intracutaneous test with 0·01 to 0·02 ml of 1 to 10 mg/ml of freeze dried extract. Large ill-defined, soft, oedematous reaction at 5 h after the test in an allergic patient with precipitins and suffering from allergic bronchopulmonary aspergillosis.

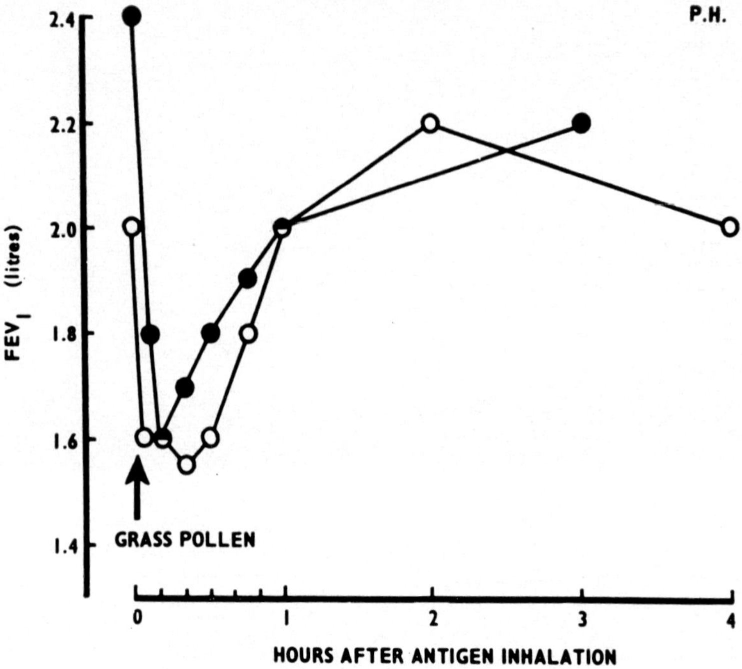

FIG. 14. Immediate, Type I, asthmatic reaction to an inhalation test, with an extract of grass pollen in an atopic subject. There is a rapid fall in the "forced expiratory volume" (FEV_1) caused by narrowing of the bronchi. The reaction resolves in 2 to 3 h, eosinophilia may occur but no fever or leucocytosis.

accompanied by a fall in ventilatory performance (Fig. 14). Non-atopic subjects with Type III sensitivity may give either a slowly developing "late" asthmatic or an alveolar reaction. Both reactions start after several hours and last about 24 h, depending on the dosage (Fig. 15). They may be accompanied by fever and leucocytosis.

Types I and III allergic reactions may occur together as in broncho-pulmonary aspergillosis in some individuals, who give an immediate asthmatic response followed later by a second delayed asthmatic reaction (Fig. 16). It is important, especially with previously untested allergens, to keep the reactions to inhalation tests within tolerable levels by carefully controlling the dose.

Serological tests. In vitro tests have only recently been developed for the allergen specific IgE reaginic antibodies mediating Type I reactions (Wide, Bennich and Johannson, 1967). Double diffusion and immunoelectrophoresis tests are well known for the demonstration of precipitins mediating Type III reactions (Figs 5, 7 and 17). Haemagglutination, complement fixation, latex fixation and other methods are also useful.

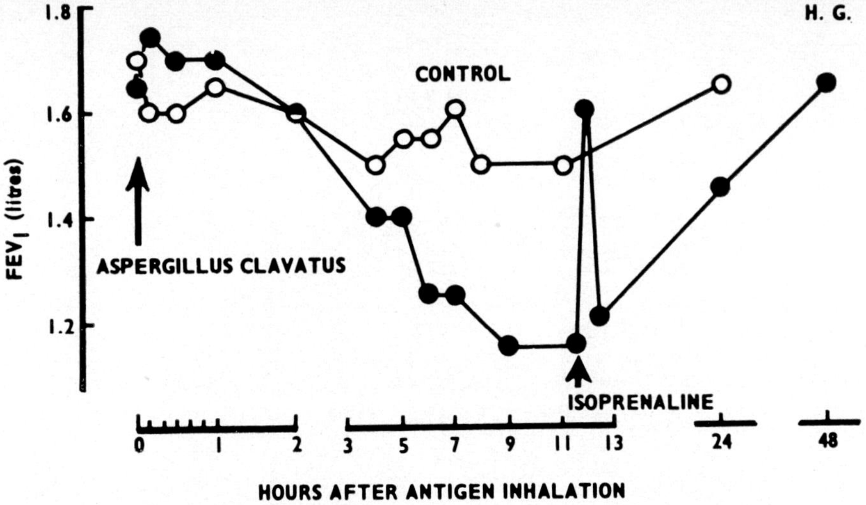

FIG. 15. Late, Type III asthmatic reaction to inhalation of *Aspergillus clavatus* spores by a non-atopic maltster with precipitins to this fungus. Slowly developing asthmatic reaction showing a fall in the "forced expiratory volume" (FEV_1) *c.* 3 h after the test, accompanied by fever and leucocytosis. Easily mistaken for an infection.

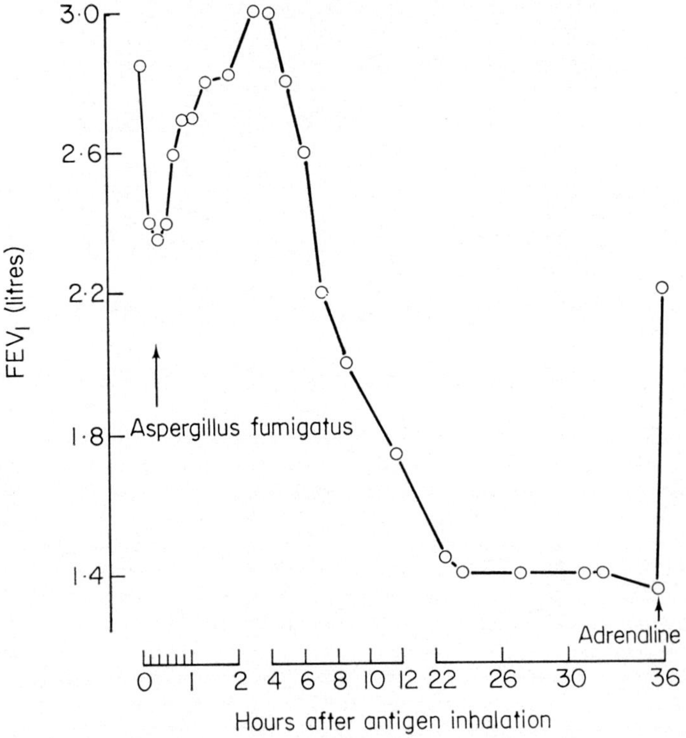

FIG. 16. Dual asthmatic reaction to inhalation of an extract of *Aspergillus fumigatus* spores by a patient with allergic bronchopulmonary aspergillosis who has developed precipitins. An immediate asthmatic reaction is followed after its resolution by a late, more prolonged and severe asthmatic reaction, accompanied by fever and leucocytosis.

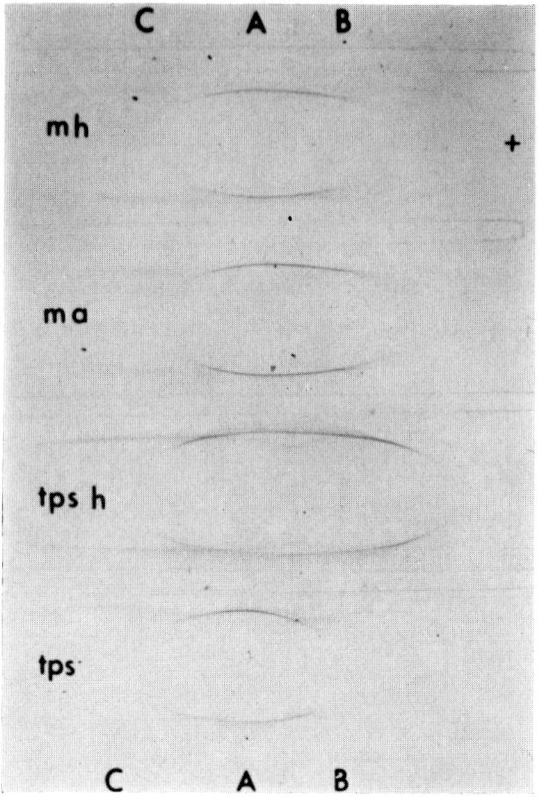

FIG. 17. Agar gel immunoelectrophoresis test for precipitins in Farmer's Lung. Test extracts were placed in wells midway between the troughs and about mid-way between the letters A and C, and were also electrophoresed. Serum from affected patients was then placed in the troughs. (mh) precipitation reactions in regions A, B and C to an extract of mouldy hay; (ma) reactions in the same regions to an extract of sterilized hay inoculated with mixed cultures of actinomycetes from mouldy hay; (tps h) reactions in the same regions to an extract of *Micropolyspora faeni* cultured on sterilized hay; (tps) reactions in the same regions to an extract of *M. faeni* grown on nutrient agar medium. Sterilized hay, heated without inoculation with *M. faeni*, did not produce any antigens. Precipitation reactions A and B are produced by negatively charged predominantly "protein" fractions, and C by a positively charged, mainly "polysaccharide" fraction. Inhalation tests with these different fractions reproduced the features of Farmer's Lung in affected subjects.

Disease Prevention

Decreasing the hazards to personnel

Descriptions in the literature of allergic respiratory diseases caused by fungus and actinomycete spores indicate the symptoms to be looked for in workers exposed to such spores and the working environments where they may be encountered. Diseases formerly classified as influenza or bronchitis, or as unavoidable hazards associated with certain occupations may now be recognized and prevented.

Mould growth in hay can be prevented by good farming practice, by adequately drying the hay before baling or stacking and by taking similar precautions before storing grain. Adding propionic or formic acids is proving useful to prevent fungi and actinomycetes growing in stored grain and silage. Mouldy hay or grain should be handled in the open air, with the handler standing to windward. If it has to be handled in an enclosed space, the handler should wear a respirator with a filter able to retain the smallest particles that otherwise would be deposited in the deepest parts of the lung and cause hypersensitivity reactions.

Where many spores are encountered in industrial operations, concentrations may be decreased by installing ventilating systems and workers at special risk should wear efficient respirators. Vertical laminar flow units (see p. 21) should be used to prevent the build-up of large concentrations of airborne allergens from materials known to carry microbial spores or antigenic products.

Serial observation for immunological changes

Regular pre- and post-employment clinical examinations should be supplemented with immunological tests to provide a baseline from which changes in immunological status can be deduced. Prick and serological tests with suitable extracts should be made before exposure and an initial serum sample should be stored for future reference.

At regular intervals prick and serological tests should be repeated. Development of skin reactivity or precipitins in a previously negative subject should be linked with a clinical check of the subject's health. Incidence and speed of sensitization should be correlated with the nature of the exposure, and where precautions are being taken against a known sensitizing agent, the development of sensitivity in initially negative subjects can serve as a measure of the efficiency of control measures.

Subjects showing immunological change and related clinical disease should avoid exposure completely. Routine use of efficient protective masks and proper disposal of contaminated clothing, or a change of employment, may be necessary. Attempts at hyposensitization, where appropriate, may sometimes help.

However, when subjects show immunological change, but no clinical manifestations, the general experience with the particular allergenic exposure will have to be used as a guide. When there has been much disease associated with exposure, contact should be avoided as soon as possible. If little or no disease has occurred, careful clinical observation, with control of exposure, may be all that is necessary.

Safety precautions must consist of the combination of frequent medical and immunological checks with a thorough examination of the air spora (Liebeskind, 1965). To prevent these diseases demands not only the realization of the possible dangers but the introduction of adequate safety procedures.

References

ANDERSEN, A. A. (1958). New sampler for the collection, sizing and enumeration of viable airborne particles. *J. Bact.*, **76,** 471.

AVILA, R. & VILLAR, T. G. (1968). Suberosis respiratory disease in cork workers. *Lancet,* **1,** 620.

BANASZAK, E. F., THIEDE, W. H. & FINK, J. N. (1970). Hypersensitivity pneumonitis due to contamination of an air conditioner. *New Engl. J. Med.*, **283,** 271.

BELIN, L., FALSEN, E., HOBORN, J. & ANDRÉ, J. (1970). Enzyme sensitization in consumers of enzyme-containing powder. *Lancet,* **2,** 1153.

BOURDILLON, R. B., LIDWELL, O. M. & THOMAS, J. C. (1941). A slit sampler for collecting and counting air-borne bacteria. *J. Hyg., Camb.*, **41,** 197.

BRINGHURST, L. T., BYRNE, R. N., & GERSHON-COHEN, J. (1959). Respiratory disease of mushroom workers: Farmer's lung. *J. Am. med. Ass.*, **171,** 15.

BUECHNER, H. A., PREVATT, A. L., THOMPSON, J. & BLITZ, O. (1958). Bagassosis: a review with further historical data, studies of pulmonary function and results of adrenal steroid therapy. *Am. J. med.*, **25,** 234.

COHEN, H. I., MERIGAN, T. C., KOSEK, J. C. & ELDRIDGE, F. (1967). Sequoiosis: A granulomatous pneumonitis associated with redwood sawdust inhalation. *Am. J. Med.*, **43,** 785.

CROFTON, J. & DOUGLAS, A. (1969). *Respiratory Diseases*. Oxford: Blackwell Scientific Publications.

DARLOW, H. M. (1969). Safety in the microbiological laboratory. *Methods in Microbiology* (J. R. Norris and D. W. Ribbons, eds) Vol. 1, p. 169. London and New York: Academic Press.

DUNNER, L., HERMON, R. & BAGNALL, D. J. T. (1946). Pneumoconiosis in dockers dealing with grain and seeds. *Br. J. Radiol.*, **19,** 506.

EMANUEL, D. A., LAWTON, B. R. & WENZEL, F. J. (1962). Maple-bark disease: Pneumonitis due to *Coniosporium corticale*. *New Engl. J. Med.*, **266,** 333.

FILIP, B. & BARBORIK, M. (1966). Bronchopulmonalni aspergiloza. *Pracovni. Lék.* **18,** 308.

FLINDT, M. L. H. (1969). Pulmonary disease due to inhalation of derivatives of *Bacillus subtilis* containing proteolytic enzyme. *Lancet,* **1,** 1177.

GREGORY, P. H. (1961). *Microbiology of the Atmosphere*. London: Leonard Hill.

GREGORY, P. H. & LACEY, M. E. (1963). Mycological examination of dust from mouldy hay associated with Farmer's Lung disease. *J. gen. Microbiol.*, **30,** 75.

GREGORY, P. H. & SREERAMULU, T. (1958). Air spora of an estuary. *Trans. Br. mycol. Soc.*, **41,** 145.

HARINGTON, J. S. (1967). Potential dangers relating to laboratory work on the aspergilli and their toxins. *S. Afr. med. J.*, **41,** 282.

HARRIS, L. H. (1939). Allergy to grain dusts and smuts. *J. Allergy*, **10,** 327.

HEARN, C. E. D. (1968). Bagassosis: An epidemiological, environmental and clinical survey. *Br. J. indust. Med.*, **25,** 267.

HIRST, J. M. (1952). An automatic volumetric spore trap. *Ann. appl. biol.*, **39,** 257.

HOŘEJŠI, M., ŠACH, J., TOMŠIKOVÁ, A. & MECL, A. (1960). A syndrome resembling farmer's lung in workers inhaling spores of aspergillus and penicillia moulds. *Thorax*, **15,** 212.

JACKSON, E. & WELCH, K. M. A. (1970). Mushroom workers' lung. *Thorax*, **25,** 25.

JAMISON, S. C. & HOPKINS, J. (1941). Bagassosis: A fungus disease of the lung. *New Orl. med. surg. J.*, **93,** 580.

JIMINEZ-DIAZ, C., LAHOZ, C. & CENTO, G. (1947). The allergens of mill dusts: asthma in millers, farmers and others. *Ann. Allergy*, **5,** 519.

LACEY, J. (1969). Spores in the air of farm buildings. *Filtration in medical and health engineering*. (K. J. Ives, M. W. Dixon, R. G. Dorman, H. M. Darlow, D. B. Purchas, R. M. Wells and W. G. Norris, eds), p. 60. Croydon: Filtration Society.

LACEY, J. (1971*a*). *Thermoactinomyces sacchari* sp.nov., a thermophilic actinomycete causing bagossosis. *J. gen. Microbiol.*, **66,** 327.

LACEY, J. (1971*b*). The microbiology of moist barley storage in unsealed silos. *Ann. appl. Biol.*, **69,** 187

LACEY, J. & LACEY, M. E. (1964). Spore concentrations in the air of farm buildings. *Trans. Brit. mycol. Soc.*, **47,** 547.

LIDWELL, O. M. (1970). Mikroorganismer: levende stof i luften. *Termisk og atmosfaerisk indeklima* (N. Jonassen, ed.) p. 481. Copenhagen: Polyteknisk Forlag.

LIEBESKIND, A. (1965). Schimmelpilzallergic in Werkstätten. *Allergie Asthma*, **11,** 62.

MAY, K. R. (1945). The cascade impactor: an instrument for sampling coarse aerosols. *J. scient. Instrum.*, **22,** 187.

MAY, K. R. (1966). Multistage liquid impinger. *Bact. Rev.*, **30,** 559.

NORTH, J. D. K. & GWYNNE, J. F. (1960). A review of possible human implications of facial eczema. *N. Z. med. J.*, **59,** 325.

ORDMAN, D. (1958). Cereal grain dusts as a cause of respiratory allergy in South Africa. *S. Afr. med. J.*, **32,** 784.

PEPYS, J. (1969). *Hypersensitivity Diseases of the lungs due to fungi and organic dusts. Monographs in Allergy*, Vol. 4. Basle: S. Karger.

PEPYS, J., HARGREAVE, F. E., LONGBOTTOM, J. L. & FAUX, J. (1969). Allergic reactions of the lungs to enzymes of *Bacillus subtilis*. *Lancet*, **1,** 1181.

PIRQUET, C. VON (1906). Allergie. *Munch. med. Wschr.*, **30,** 1457.

RADONIC, M. (1966). Systemic allergic reactions due to occupational inhalation of tuberculin aerosols. *Ind. Med. Surg.*, **35,** 24.

RAMAZZINI, B. (1713). *De morbis artificum diatriba*. Translated by W. C. Wright (1964). New York: Hafner.

RIDDLE, H. F. V., CHANNELL, S., BLYTH, W., WEIR, D. M., LLOYD, M., AMOS, W. M. G. & GRANT, I. W. B. (1968). Allergic alveolitis in a maltworker. *Thorax*, **23,** 271.

SAKULA, A. (1967). Mushroom worker's lung. *Brit. med. J.*, **3,** 708.

SALVAGGIO, J. E., BUECHNER, H. A., SEABURY, J. H. & ARQUEMBOURG, P. (1966). Bagassosis: 1. precipitins against extracts of crude bagasse in the serum of patients. *Ann. intern. Med.*, **64,** 748.

SCHILLING, R. S. F., HUGHES, J. P. W., DINGWALD-FORDYCE, I. & GILSON, J. C. (1955). An epidemiological study of Byssinosis among Lancashire cotton workers. *Br. J. indust. med.*, **12,** 217.

SKOULAS, A., WILLIAMS, N. & MERRIMAN, J. E. (1964). Exposure to grain dust. II. A clinical study of the effects. *J. occup. Med.*, **6,** 359.

TOWEY, J. W., SWEANEY, H. C. & HURON, W. R. (1932). Severe bronchial asthma apparently due to fungus spores found in maple bark. *J. Am. med. Ass.*, **99,** 453.

VALLERY-RADOT, P. & GIROUD, P. (1928). Sporomycose des pelleteurs de grains. *Bull. Mém. Soc. méd. Hôp. Paris*, **52,** 1632.

WEDUM, A. G. & KRUSE, R. H. (1969). *Assessment of risk of human infection in the microbiological laboratory*. Miscellaneous Publication 30. Fort Detrich, Maryland, U.S.A.: Department of the Army.

WENZEL, F. J. & EMANUEL, D. A. (1967). The epidemiology of maple bark disease. *Archs. envir. Hlth.*, **14,** 385.

WIDE, L., BENNICH, H. & JOHANNSON, S. G. O. (1967). Diagnosis of allergy by an *in-vitro* test for allergen antibodies. *Lancet*, **2,** 1105.

WILLIAMS, N., SKOULAS, A. & MERRIMAN, J. E. (1964). Exposure to grain dust. I. A survey of the effects. *J. occup. Med.*, **6,** 319.

WOLF, F. T. (1969). Observations on an outbreak of pulmonary aspergillosis. *Mycopath. mycol. appl.*, **38,** 359.

Carcinogenic Hazards in the Microbiology Laboratory

J. M. Wood and R. Spencer*

The British Food Manufacturing Industrial Research Association, Randalls Road, Leatherhead, England

Over the last decade, our growing awareness of hazards involved in working with certain strong carcinogens has resulted in legislation banning or strictly controlling the use of these compounds in factories (Anon, 1967). These regulations, however, do not apply to laboratories where workers may not be aware of the carcinogenicity of reagents which they are using. Microbiologists, particularly when working with pathogenic organisms, face hazards which other laboratory workers do not. This sometimes overshadows the fact that in the safety context, the microbiology laboratory is a chemical laboratory with additional biological hazards and microbiologists must be aware of both types of hazard. At least three techniques which in the past have been commonly used in microbiological laboratories involve the use of strong chemical carcinogens (Table 1).

Napthylamines

The ability of an organism to reduce nitrate to nitrite is a commonly used diagnostic test. The detection of nitrite is carried out by providing conditions under which available nitrite enters into a diazotization reaction and is

TABLE 1. Proven carcinogens used in microbiology laboratories

Compound	Use
Napthylamines	Nitrate reduction tests
Benzidine	Detection of hydrogen peroxide and bacterial cytochromes
β-propiolactone	Sterilant
Isoniazid (isonicotinic acid hydrazide)	Tuberculosis sensitivity test
Aflatoxin	Investigations of growth and production Assay systems

* Present address: J. Sainsbury Ltd., Stamford House, Stamford Street, Blackfriars, London, S.E.1., England.

revealed by the development of a red colour. The reagents are normally sulphanilic acid and a napthol derivative. α-napthylamine or dimethyl α-napthylamine have been commonly used and are listed recently in a standard identification text as nitrite test reagents (Cowan and Steel, 1965). The Carcinogenic Substances Regulations (Anon, 1967) recognize the hazard of β-napthylamine and its salts as well as that of α-napthylamine. The substitution of α-napthol for α-napthylamine (Spencer, 1969) is less hazardous.

Benzidine

Aromatic amines are the outstanding carcinogens from the laboratory hazard aspect. Benzidine has been used widely in various types of laboratory, e.g. in the detection of blood, the analysis of water and milk and the determination of sulphates and metals. Whilst exposure in the laboratory might be expected to be small compared to that in the industrial situation, recent evidence suggests that laboratory exposure to carcinogenic amines has actually caused cases of bladder cancer (Case, 1966, 1967). The carcinogenic hazard of benzidine is now widely recognized and the manufacture, presence and use of benzidine or its salts are prohibited in factories.

Benzidine has commonly been used, and is still possibly in use, in two microbiological tests. The production of hydrogen peroxide by certain anaerobes, e.g. *Clostridium oedematiens* and *Clostridium botulinum*, is particularly well demonstrated by growth on heated blood agar to which benzidine has been added (Gordon and McLeod, 1940). A similar medium, Benzidine Erythrocytle Brain Heart agar, has been used for many aerobic organisms (Kraus *et al.*, 1957). M-amino acetanilide may be substituted for benzidine but is inferior as an indicator. The use of pyrolusite (manganese dioxide) agar (Kneteman, 1947) offers the best safe alternative to the use of benzidine for hydrogen peroxide detection. Manganese dioxide layered in agar over the surface of an agar plate renders the medium opaque. Hydrogen peroxide causes a clearing of the opacity indicating hydrogen peroxide porducing colonies. The second method utilizes a benzidine salt, benzidine dihydrochloride, in the detection of bacterial cytochromes (Deibel and Evans, 1960). Benzidine dihydrochloride and hydrogen peroxide react with iron porphyrins to give a blue coloration due to quinodic bond formation. In this method plates are flooded with the benzidine reagent, creating a considerable exposure hazard. We have found no simple alternative to this method for screening large numbers of organisms, the normal chemical methods of extraction and spectroscopic analysis are lengthy but safer. If the benzidine method is employed, careful precautions should be taken to avoid contact with the skin or inhalation of vapour.

β-propiolactone

β-propiolactone is a biological alkylating agent which has been used as sterilant in both industry and the laboratory. It is more effective than other similar compounds but it has been found to be carcinogenic. Alkylating agents are not among the strongest carcinogens and there is little evidence linking them with human cancer. However, the use of β-propiolactone in both industrial and laboratory situations requires caution in respect of its carcinogenic nature.

Isoniazid (isonicotinic acid hydrazide)

This drug, which is used in the treatment of tuberculosis, has produced lung tumours in some mice. It is used in medical microbiology laboratories in sensitivity tests for this organism and whilst much sensitivity media is now centrally prepared, some laboratories may still produce their own supplies. Although in the treatment of tuberculosis, the benefits of the drug can be considered to outweigh the slight carcinogenicity hazard, laboratory workers have no reason to accept any extra risk and should act accordingly.

Most microbiologists are well aware of the biological hazards of their own work but one group of extremely potent carcinogens are worthy of extra note.

The aflatoxins

This group of complex lactones produced by *Aspergillus flavus* are at present the strongest carcinogens known. They will produce tumours in rats when fed at 1 mg/day and are associated with acute and fatal disease in humans (Shank *et al.*, 1971). Because of their extreme toxicity and carcinogenicity for warm blooded animals, extreme care should be taken in the handling of both toxigenic cultures and any material which may be contaminated with toxin. The use of respirator masks and disposable laboratory coats and gloves is recommended when working with dry preparations and stringent precautions must be taken to monitor and decontaminate all working areas. Anyone contemplating work in the mycotoxin field is strongly advised to read the introduction to Goldblatt's (1969) book "Aflatoxin", in which he describes precautionary measures in dealing with these carcinogens, and also the Chapter on the implications of fungal toxicity to human health.

Selenium

It is appropriate here to mention a suggested biohazard which may be particularly relevant to microbiological laboratories. Sodium selenite which is commonly used in media for the isolation of salmonellae is known to be toxic to humans. It has recently been suggested (Robertson, 1970) that this compound may also produce teratogenic effects (i.e. abnormalities of the developing foctus) in pregnant laboratory workers. Whilst this has by no means been proved, it would seem reasonable for women of child-bearing age to avoid inhalation of selenite containing powder, which becomes air-borne during the handling of dehydrated media. At least one manufacturer, Oxoid Ltd., now supplies selenite media as powdered broth and crystalline sodium bi-selenite which are dissolved separately and mixed as solutions. This reduces the hazard of selenite inhalation to the level associated with normal chemical manipulations with selenite compounds.

Whilst chemical carcinogens have been known for a considerable time, little information on their identity and hazards has become widely available to microbiologists or indeed to the majority of laboratory workers. This information, if gathered at all, is accumulated slowly. Chemical carcinogenesis is a slow process roughly related to the life span of the species concerned, and avoidance of exposure is therefore particularly important where young workers are concerned. The lack of formal instruction in these hazards, combined with the insidiously slow response time in humans (5–40 years) places a heavy responsibility on senior microbiologists supervising young colleagues and laboratory staff.

References

ANON. (1967). The Carcinogenic Substances Regulations. Pub. No. 879. London: H.M.S.O.

CASE, R. A. M. (1966; 1967). *A. Rep. Br. Emp. Cancer Campaign Res.*, **44,** 56; **45,** 90.

COWAN, S. T. & STEEL, K. J. (eds) (1965). The Identification of Medical Bacteria. Cambridge: Cambridge University Press.

DEIBEL, R. H. & EVANS, J. B. (1960). Modified benzidine test for the detection of cytochrome containing respiratory systems in microorganisms. *J. Bact.*, **79,** 356.

GOLDBLATT, L. A. (1969). *Aflatoxin.* New York and London: Academic Press.

GORDON, J. & MCLEOD, J. W. (1940). Distinguishing *Clostridium novyi* from other bacteria associated with gas gangrene. *J. Path. Bact.*, **50,** 167.

KNETEMAN, A. (1947). A reaction on the formation of hydrogen peroxide by microorganisms in solid media. *Antonie van Leeuwenhoek*, **13,** 55.

KRAUS, F. W., NICKERSON, J, F. PERRY, W. I. & WALKER, A. P. (1957). Peroxide and peroxidogenic bacteria in human saliva. *J. Bact.*, **73,** 727.

ROBERTSON, D. S. F. (1970). Selenium—a possible teratogen. *Lancet*, 518.

SHANK, R. C., BOURGEOIS, C. H., KESCHAMRAS, N. & CHANDAVINOL, P. (1971). Aflatoxin in autopsy specimens from their children with an acute disease of unknown aetiology. *BIBRA Inf. Bull.*, **10,** (4) XXVI.

SPENCER, R. (1969). New procedure for determining the ability of microorganisms to reduce nitrate and nitrite. *Lab. Pract.*, **18,** 1286.

Safety in the Use of Radioactive Isotopes*

J. N. ANDREWS AND D. J. HORNSEY

Centre for Nuclear Studies, University of Bath, Bath, BA2 7AY, Somerset, England

Isotope, or tracer, techniques have contributed greatly to the development of the biological sciences. Hevesy (1962) carried out work with the natural radioelements in the early years of this century. At this time, work was limited because of the lack of suitable tracers for the biologically important elements especially carbon and hydrogen. The discovery of the stable isotope of hydrogen, deuterium, was followed by its use for labelling molecules in tracer experiments in the 1930's; deuterium labelled molecules being detected by mass spectrometry. Radioactive tracers for hydrogen, carbon, sulphur, phosphorus and many other elements became readily available after 1945 and a very large number of compounds labelled with such tracers are now produced. Tracer techniques became greatly facilitated because of the ease with which radioactive materials may be detected. The use of radioactive tracers, however, involves a health hazard due to the exposure of the experimenter to the radiations emitted by the radioisotope, either as a source external to the body or as an internal source if the material is ingested. Before discussing these hazards and how to control them in more detail we shall outline briefly some of the fundamental properties of radioisotopes.

Properties of Radioisotopes

Radioactive isotopes are those whose nuclei are unstable. They become stable by emitting either α-particles (helium nuclei), β-particles (energetic electrons) or γ-radiation (high energy electromagnetic radiation). The emission of γ-radiation is also very frequently associated with α- or β-particle emission. Natural radioisotopes occur amongst the heavy elements, from lead to uranium, and they decay by either α- or β-particle emission. Radioisotopes of the lighter elements do not generally occur naturally, but can be prepared by a variety of nuclear reactions. They generally decay by

* Some Acts and Regulations for the U.K. are given on p. 211.

β-particle emission, positron (positive electron) emission or by the nuclear capture of a K-shell orbital electron (EC). Isotopes which decay by β-particle emission have too many neutrons in the nucleus for stability and those which decay by positron emission or K-electron capture have too few neutrons for stability. The preparation of radioisotopes by the neutron capture reaction in which a stable nucleus captures a neutron and generally becomes radioactive, can be easily effected by irradiating a sample of the stable element with neutrons in a nuclear reactor. The use of such nuclear reactions has enabled radioisotopes of most of the elements to be produced.

Decay rate, activity and half-life

The rate of decay, or activity, of a radioisotope is given by:

$$\frac{dN}{dt} = -\lambda N \qquad (1)$$

where λ is the radioactive decay constant for the isotope concerned and N is the number of atoms of the isotope present in the sample. This equation may be expressed in the integrated form:

$$N_t = N_o \, e^{-\lambda t} \qquad (2)$$

from which

$$\log_e N_t = \log_e N_o - \lambda t \qquad (3)$$

and on converting $\log_e$ to $\log_{10}$:

$$2{\cdot}303 \log_{10} N_t = 2{\cdot}303 \log_{10} N_o - \lambda t \qquad (4)$$

where N_t is the number of atoms of the isotope present at time t and N_0 is the number of atoms initially present. The half-life, $t_{\frac{1}{2}}$, of a radioisotope is defined as the time required for half of the atoms initially present to decay. Substituting $N_t = N_o/2$ and $t = t_{\frac{1}{2}}$ in equation (4) shows that half-life and the radioactive decay constant are related as follows:

$$0{\cdot}693 = \lambda t_{\frac{1}{2}} \qquad (5)$$

Since equation (1) shows that the decay rate or activity of a radioisotope is proportional to the number of atoms of it present, equation (4) may be re-written as:

$$2{\cdot}303 \log_{10} A_t = 2{\cdot}303 \log_{10} A_o - \lambda t \qquad (6)$$

where A_t and A_o are the activities of the radioisotope sample at times t and zero, respectively.

Units of radioactivity

The unit of radioactivity is defined as that quantity of a radioisotope which has an activity of $3{\cdot}7 \times 10^{10}$ disintegrations/sec. This unit is called the curie (Ci) and common sub-units are the milli-curie (mCi) or 10^{-6}Ci, the microcurie (μCi) or 10^{-6}Ci, and the nanocurie (nCi) or 10^{-9}Ci.

Specific activities

The ratio of the activity of a radioisotope sample to the mass of the element or labelled compound present is called the **specific activity**. A radioactive preparation which has no stable isotopes of the element present is termed "carrier free". The specific activities of carrier free radioisotope preparations may be calculated using equation (1). Such specific activities are dependent upon the value of λ or the half-life of the isotope concerned and are generally very high, being of the order of several thousand Ci/g for half-lives of the order of a few weeks.

The energy of the radiation emitted in nuclear decay

The energies of the radiations emitted by a radioisotope are generally characteristic of that isotope. These energies are expressed in millions of electron volts (MeV); the basic unit, the electron volt (eV) being the amount of energy acquired by an electron on being accelerated by a potential difference of 1 V. The electron volt is a minute amount of energy and a 1 kW electric fire produces $2{\cdot}25 \times 10^{19}$ MeV of energy/h. The energies of the nuclear radiations associated with radioisotope decay are generally of the order of MeV. Although this is a very small amount of energy, it is a considerable amount in molecular terms and if absorbed by matter it may cause the destruction or alteration of a very large number of molecules. As we shall see later, the concept of radiation dose is in effect the estimation of the energy deposited in matter by nuclear radiations.

The heavy radio-elements emit α-particles which are mono-energetic and typically have an energy of *c.* 5 MeV. Because of their comparatively great mass, these particles are absorbed very readily in matter and deposit their energy within a short distance.

The β-emitting radioisotopes emit β-particles (or positrons) with a maximum energy which is characteristic of the radioisotope. These maximum energies range from *c.* 0·018 MeV for hydrogen-3 (tritium) to several MeV. For example, yttrium-90, the decay product of strontium-90, (Table 1) has an energy of 2·27 MeV. Although each radioisotope has a characteristic maximum β-particle energy, most of the β-particles emitted have energies less than this maximum, ranging from zero up to the maximum.

The energies of the γ-rays emitted by radioisotopes are characteristic of the isotopes. The numbers and intensities of the γ-rays emitted are

TABLE 1. Physical data* on radioisotopes in common use in the biological laboratory

Radioisotope	Half-life	Principal radiation emitted	Energy (MeV)	Specific γ-ray constant (Γ) R.mCi^{-1}.h^{-1}cm^2	Toxicity when used as solution
Caesium-137 (Cs-137)	30 years	β^-	0·51	—	High
		γ	0·662	3·3	
Calcium-45 (Ca-45)	165 days	β^-	0·254	—	High
Carbon-14 (C-14)	5760 years	β^-	0·16	—	Moderate
Chlorine-36 (Cl-36)	3×10^5 years	β^-	0·714	—	High
Chromium-51 (Cr-51)	27·8 days	γ	0·323	0·16	Moderate
Cobalt-60 (Co-60)	5·26 years	β^-	0·31	—	High
		γ	1·17, 1·33	13·2	
Hydrogen-3 (tritium)	12·26 years	β^-	0·018	—	Low
Iodine-131 (I-131)	8 days	β^-	0·61	—	High
		γ	0·36	2·2	
Iron-59 (Fe-59)	45 days	β^-	0·27, 0·46	—	Moderate
		γ	1·10, 1·29	6·4	
Phosphorus-32 (P-32)	14·3 days	β^-	1·71	—	Moderate
Sodium-22 (Na-22)	2·6 years	β^+	0·54	—	High
		γ	0·51, 1·28	12·0	
Sodium-24 (Na-24)	15 h	β^-	1·39	—	Moderate
		γ	1·37, 2·75	18·4	
Strontium-90 (Sr-90) (Yttrium-90)	28 years	β^-	0·54, 2·27	—	V. High
Sulphur-35 (S-35)	87·2 days	β^-	0·167	—	Moderate
Zinc-65 (Zn-65)	245 days	β^+	0·325	—	Moderate
		γ	0·51, 1·11	2·7	

* Taken from the Radiochemical Manual (Wilson, 1966).

determined by the complexity of the nuclear energy levels for the decay product formed on α- or β-decay or K-electron capture.

The decay properties of some radioisotopes commonly used in biological work are listed in Table 1 and a complete list may be found in the Radiochemical Manual (Wilson, 1966) from which the data of Table 1 were taken.

Some Applications of Radioactive Tracers in Biology

Typical applications of tracer techniques to biological problems may be enumerated thus:

1. Metabolic studies and enzyme assay involving the use of labelled substrates (Oldham, 1968).
2. Determination of trace levels of amino acids, sugars and steroids in fluids by radioisotope dilution analysis (Gorsuch, 1968*a*, *b*).
3. Determination of trace elements by radioactivation analysis (Bowen and Gibbons, 1963).

All these generally require the ultimate determination of small amounts (nCi–μCi) of radioactivity. They may, however, also involve handling larger amounts of activity at early stages in the experiments. The determination of amino acids may, for example, be effected by determining the activity of the complexes formed with a copper-64 labelled reagent. The preparation of the reagent involves handling activities of the order of several mCi, but the final activity measurement for the determination of a few μg of an amino acid, will involve activities of less than 1 μCi. Another example, in which large activities may initially be handled, is the determination of trace elements by radioactivation. In this procedure the material to be analysed is irradiated by neutrons in a nuclear reactor and the major elements of the matrix as well as the trace elements are generally made radioactive. Most biological materials produce high activities due to activation of sodium, potassium and phosphorus. After the irradiation it is generally necessary to chemically separate the low levels of activity (less than 1 μCi) associated with the trace elements from the high levels of activity (several to 100 mCi) associated with the matrix. Most of the activity present is unwanted and must end up as radioactive waste.

Radiation Hazards

Radiations are absorbed in matter by processes which involve ionization and excitation of the atoms or molecules of the absorbing medium. This property is made use of in radiation detectors which enable nuclear radiations to be measured by the ionization which they cause in a gas, the fluorescence which they cause in a phosphor, or the charge carriers produced in a semiconductor device. All these processes involve the deposition of energy in the absorbing material. If nuclear radiations are absorbed in biological materials, the energy deposited may cause the destruction or alteration of chemical molecules. If the chemicals involved are enzymes, nucleic acids or other vital cell molecules, then such changes may result in either the death of the cell or the induction of mutations.

The radiation dose to a material which absorbs nuclear radiations may be defined in terms of the energy deposited in the material. The unit is the rad which is an exposure to radiation resulting in an energy absorption of

$\frac{1}{100}$ joule/kg (or 100 ergs/g).

The extent of biological damage for a given dose measured in rads may vary for different kinds of radiations. Another unit of dose which takes into account this variation in biological damage is therefore used in radiation protection. This unit is the rem and the dose in rems is related to the dose in rads by:

$$\text{dose in rems} = \text{dose in rads} \times \text{quality factor}$$

The quality factor relates the effectiveness of the radiation for causing biological damage to that of X- or γ-radiation. For X- and γ-rays, and β-particles the value of the quality factor is unity, for α-particles it is 10. Another unit of radiation dose which is often used to describe the intensity of a radiation field is the roentgen (R). The roentgen, being a measure of the radiation field intensity, is an expression of the "exposure dose", whereas the rad or rem being measures of energy absorption, are referred to as units of "absorbed dose". For most practical purposes we may regard the roentgen, the rad and the rem as being the same for exposure to β- and γ-emitting sources.

Radiation hazards can arise due to radioisotope sources which are external to the body and also due to radioisotopes which have been ingested and are distributed within the body. The former are referred to as **external hazards** and the latter as **internal hazards**. External hazards may be present in handling radioactive materials which are either **closed** or **open** sources. A closed source is one in which the radioactive material is permanently sealed within a containment capsule, whereas an open source is one in which the material is present, generally in solution, in an open chemical system. In the case of open sources it is clearly possible to ingest the radioactive material and so present an internal as well as an external hazard.

External hazards may arise with β- and γ-emitting sources and are generally significant with mCi but not of great consequence for μCi level sources. Internal hazards arise in handling all open sources and may be considerable even with μCi level sources, depending upon the isotope involved. Even if the isotope is one of low radiochemical toxicity such as tritium (Table 1), care must be exercised over long periods to ensure that chronic ingestion of small amounts of activity does not lead to a significant build-up of ingested material.

Estimation of external hazards from radioactive sources

Since α-particles are absorbed by a few centimetres of air or by the outer layers of the skin, they do not present an external radiation hazard. External hazards only arise with γ-ray or β-particle emitting materials. The dose rate due to a γ-emitting source may be estimated from the formula:

$$D = \frac{10^3 A.\Gamma}{d^2} \text{ mR/h} \tag{7}$$

where A is the source activity, mCi, Γ is the specific γ-ray constant in R.mCi^{-1}h^{-1}cm^2, and d is the distance from the source in cm.

Values of the specific γ-ray constant in the units quoted may be found in the Radiochemical Manual (Wilson, 1966) or Table 1.

For a β-emitting source, the assessment of the dose rate by calculation is more difficult because of the energy spectrum of the β-particles emitted and because of the air absorption or self-absorption of the β-particles within the source. The following formula gives the approximate dose rate for a β-emitting source of maximum β-energy greater than 1·5 MeV at distances up to 30 cm:

$$D = 10^3 A/3d^2 \text{ mR/h} \tag{8}$$

where A is the source activity, μCi, and d is the distance (cm) from the source.

Although dose rates may be estimated from formulae such as those above, if work is undertaken which involves more than occasional external exposures at significant levels, the dose rates involved should be measured with a dose-rate meter (Fig. 1).

Maximum permissible levels of radiation exposure

Exposure of the general body tissues to ionizing radiation will result in damage which can only effect the exposed person; such effects are referred to as **somatic**. Exposure involving the germ cells, however, can induce mutations which may be transmitted to the exposed person's descendants. Such **genetic** effects if widespread within a population may lead to a genetic deterioration of that population. An upper limit to the radiation

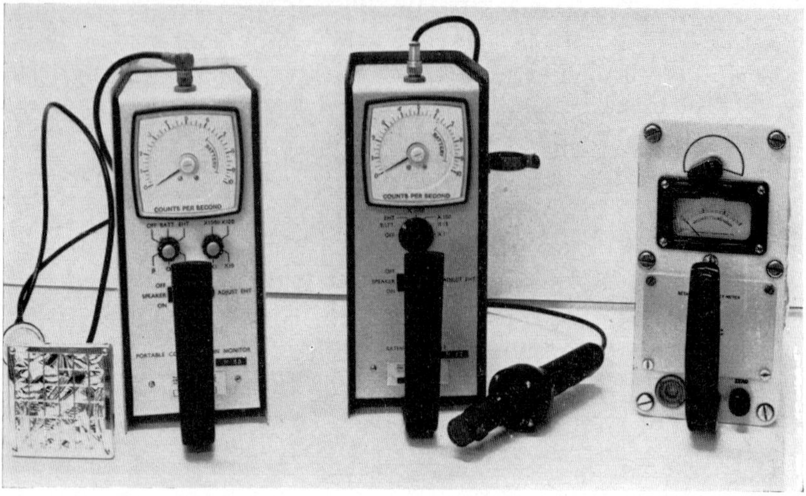

FIG. 1. Contamination monitors with a scintillation detector (left), a Geiger-Müller detector (centre), and a dose-rate meter (right).

dose, which may be received by the gonads or the whole body, would be one which limited the somatic and genetic consequences to acceptable levels. Limits for radiation exposure of the whole body and various organs have been recommended by the International Commission on Radiological Protection (Anon, 1966).

The recommended Maximum Permissible Level (MPL) for radiation exposure of the whole body is given by the formula:

$$D = 5\,(N\text{–}18)\text{ rems} \qquad (9)$$

where D is the accumulated dose and N is the age (years) of the subject. Broadly, this formula implies that there should be no occupational radiation exposure before the age of 18, and that the average dose after age 18 should not be greater than 5 rems/year. There are further limitations on the maximum single dose which is permissible, but for most tracer work carried out regularly throughout the year, the weekly dose should be limited to less than 100 mrems/week for whole body exposure. This value is then the limit for a genetically significant radiation dose.

Radioisotopes which are ingested will become distributed amongst the various body organs according to their biochemical behaviour. MPL of dose to which various body organs may be subjected have been recommended by the International Commission for Radiological Protection (ICRP) (Anon, 1966). The recommended MPL for organs other than the gonads and bone marrow is 300 mrems/week. The organ which is likely to suffer most biological damage due to the ingestion of a particular radioisotope is called the **critical organ**. The activity of the radioisotope which must be present in the body to contribute the maximum permissible dose to the critical organ is called the **body burden** for that radioisotope.

The maximum permissible body burdens of some radioisotopes are quoted in Table 2. These values are taken from the report (Anon, 1959) of the ICRP and a complete list may be found in this reference. It may be noted that the body burden for iodine-131 labelled iodide is very low. This is because iodide is taken up selectively by the thyroid where it contributes a high absorbed dose. The value for strontium-90 is similarly low because of its selective incorporation into bone. In comparison the levels for tritium and carbon-14, when ingested in elemental form, or as water or carbon dioxide, are high because of their general distribution throughout the body.

TABLE 2. Maximum permissible body burdens for some radioisotopes*

Isotope	Critical organ	Maximum permissible burden in total body (μCi)
$^{3}_{1}H$ (HTO or $H^{3}{}_{2}O$)	Body tissue	10^3
$^{14}_{6}C$ (CO_2)	Fat	300
$^{22}_{11}Na$ (soluble)	Total body	10
$^{24}_{11}Na$ (soluble)	Total body	7
$^{32}_{15}P$ (soluble)	Bone	6
$^{35}_{16}S$ (soluble)	Testis	90
$^{36}_{17}Cl$ (soluble)	Total body	80
$^{59}_{26}Fe$ (soluble)	Spleen	20
$^{60}_{27}Co$ (soluble)	Total body	10
	Total body	60
$^{65}_{30}Zn$ (soluble)	Prostate	70
	Liver	80
$^{90}_{38}Sr$ (soluble)	Bone	2
$^{131}_{53}I$ (soluble)	Thyroid	0·7

* Data taken from Recommendations of the ICRP Report of Committee II (Anon, 1959).

Radiation Protection from External Hazards

Armed with a knowledge of the probable dose-rate from a particular radioactive source, several courses are open to an operator in order that the dose received may be kept as low as possible.

1. He may increase the air distance between the source and his body by using some form of remote handling device. Such a procedure takes into account the Inverse Square Law which states that the intensity of the

radiation (I) is inversely proportional to the square of the distance (d) from the source:

$$I \propto \frac{1}{d^2} \tag{10}$$

The dose rate formulae already quoted have taken this factor into account. It should be noted that even for very small activities (a few μCi) the dose rate close to the source or in contact with it, can be considerable for both γ-ray and β-emitting isotopes. However, at only a few cm distance, the dose rate from such small activities becomes very low. The use of small tongs or dissecting forceps to handle them therefore results in a considerable reduction of the tissue dose to the fingers. With higher activities (mCi–Ci), specially designed handling devices 3 ft or more in length may be necessary (Fig. 2). It is good laboratory practice to handle all small sources with forceps but the need for long remote handling devices is only occasional in a tracer laboratory.

FIG. 2. Using remote handling devices for millicurie levels of radioactivity. The unpacking of material from a reactor irradiation can.

2. Remote handling devices are cumbersome to use so the operator may decide to make use of shielding as a protective measure. By placing some absorber material between the source and the operator the dose received may be reduced to very low levels. Table 3 gives typical shields for the various types of radiation.

TABLE 3. Shielding required for ionizing radiation

Radiation	Typical energy range	Penetration in air	Shielding required
alpha (α)	5–10 MeV	2–8 cm	α-particles are stopped by a piece of paper.
beta (β)	0·018–5 MeV	20–1000 cm	Aluminium, walls of containing vessels, Perspex. 3mm Perspex will stop β-particles of 1 MeV and 2·5 cm will stop particles of 4 MeV.
gamma (γ)	0·05–2·5 MeV		Lead, concrete. Half-value layer (thickness to reduce intensity to half) of lead for Co–60, 1·6 cm; Cs–137, 0·8 cm. Tenth value layer of lead for Co–60, 4·5 cm; Cs–137, 2·2 cm. Half-value layer of concrete for Co–60, 12 cm; Cs–137, 9 cm. Tenth value layer of concrete for Co–60, 32 cm; Cs–137, 22 cm

3. Shielding, of course, is also a cumbersome procedure and reduces the speed of operations. It may therefore be more convenient to conduct the work as rapidly as possible and dispense with shielding. With a knowledge of the approximate dose from the source and a trial run with a dummy source, the time spent in its vicinity to conduct the experiment may be determined and the approximate dose calculated. Such information will tell the operator the feasibility of conducting the experiment without remote handling devices or shields by limiting the dose he receives through reduction of the time spent on the operation.

Radiation Protection from Internal Hazards

(*arising in the use of β, γ-emitting sources*)

Since the α-emitters are generally confined to the elements heavier than lead, their application to biological problems is very limited. The internal hazards in using open sources of α-emitting isotopes are very great and since the isotopes used as tracers for biological and bio-chemical problems are the β, γ-emitting isotopes of elements lighter than lead, this discussion of radiation protection is confined to the use of these materials.

By far the greatest hazard in the laboratory using open sources of β, γ-emitting isotopes is the problem of personal contamination by and possible ingestion of radioactive materials. Ingestion of radioactive material may result from breathing contaminated laboratory air or by contamination of the mouth, or by accidental entry through cuts and wounds in the skin. Air contamination can arise not only from radioactive gaseous materials but from dust borne contamination so that clean laboratory conditions are an aid to reducing entry by this route. To minimize the possibility of personal contamination, contamination of laboratory surfaces and equipment must be controlled. Apart from protection of health, it is also essential to limit the spread of contamination in a laboratory because it could lead to spurious results in further tracer work. The following series of precautions which must be observed in a tracer laboratory are very similar to the normal precautions for work in a bacteriological laboratory.

Precautions to be observed in a laboratory using open sources of β, γ-emitting isotopes

1. Mouth operations are not allowed in the laboratory: **no** pipetting by mouth, **no** smoking, eating or drinking, **no** chewing of pencils or licking labels, **no** applying cosmetics and **no** mouth glass-blowing.
2. Protective clothing **must** be worn, the very minimum being a laboratory coat and either rubber or disposable plastic gloves. The laboratory coat should be worn in the laboratory at all times and the gloves when handling open radioactive sources.
3. Work **must** be conducted in a confined area such that, if there is a spill of labelled material, the area of contamination is known and is contained. Such confined areas should be in fume-cupboards when using radioactive gases and powders, and on the bench in large paper lined trays for liquids (Fig. 3).
4. All radioactive liquids **must** be double contained. For this purpose, vessels containing radioactive solutions must be placed in containers packed with absorbing materials such as vermiculite or cotton wool. In the event of breakage of the original vessel, contamination will be limited and the spill contained.
5. Protective gloves **must** not be worn when operating light switches, counting and monitoring equipment or opening drawers. If it is inconvenient to remove gloves for such operations, paper tissues should be used to touch the switches, etc. Such a procedure prevents the spread of contamination.
6. On removing or putting on gloves, particularly those that have been

FIG. 3. Laboratory area prepared for work with open sources of radioisotopes.

used several times, it is important not to touch the outside of the gloves with the hands.

7. Hands **must** be washed and monitored before leaving the laboratory. It is important to realize that radioactive contamination cannot be neutralized. It can only be placed in a safe place and allowed to decay.

The suitability of laboratories for radioactive work

Laboratory areas suitable for radioactive work may be classified as: low level (Class C), medium level (Class B) and high level (Class A), according to the activity and toxicity of the radioisotopes to be used (Table 1). Table 4 shows the toxicity and activity of isotopes that may be used in these categories of laboratory. A Class A laboratory has no real maximum level of activity. Some modifying factors to be applied to these levels for different manipulations are also shown in Table 4.

For most work in biology, a Class C laboratory is all that is required (Fig. 3). This is a normal chemical laboratory with some simple additions. These include easily cleaned non-absorbing bench tops and floors covered with continuous strips of linoleum or welded vinyl sheets. Formica is a very satisfactory material for bench tops but wooden benches may be covered with polythene sheets to make a suitable alternative. Cracked bench tops make decontamination almost impossible. Cracks in the flooring must also be covered with adhesive tape to avoid contamination in the

TABLE 4. Maximum quantities* of radioisotopes which may be used in the different classes of radiochemical laboratory

Toxicity class of radioisotope	Maximum activity which may be used in laboratory of class: A	B	C
Very High	10 mCi+	1 mCi	10 μCi
High	100 mCi+	10 mCi	100 μCi
Moderate	1 Ci+	100 mCi	1 mCi
Low	10 Ci+	1 Ci	10 mCi

Modifying factors to be applied to the above quantities for various procedures

Procedure	Modifying factor
Storage	X 100
Simple wet operations (dispensing, transferring, etc.)	X 10
Simple chemical reactions	X 1
Complex operation with high spill risk	X 0·1
Simple dry operations	X 0·1
Dry, dusty operations	X 0·01

* Data taken from Sherwood (1959).

event of spills. For a Class C laboratory, it is preferable but not essential for the walls and ceilings to be of a smooth gloss finish. A Class B laboratory would certainly require such a finish. If work is contemplated which involves the production of radioactive gases or possible volatilization of radioactive materials, it is essential that the laboratory should also have an efficient fume hood in which this work may be carried out.

Chemical operations with radioactive materials must be performed over a contained area. This may be in the form of a large enamelled or plastic drip tray about 18 × 24 × 2 in. (Fig. 3). The tray should be lined with absorbent paper coated with polythene (Whatman Benchcote) on its underside, so that spilt liquids are absorbed and do not pass through to contaminate the tray. Only radioactive materials and contaminated apparatus should be kept in the tray and all non-active chemicals and clean apparatus kept outside. As well as limiting the possible spread of contamination the tray also acts as a useful boundary for distinguishing contaminated and non-contaminated apparatus. It is useful to have containers in the tray for holding active liquid and solid waste as it accumulates during an experiment. Normal laboratory glassware is suitable for conducting radioactive work but, because of the need to obviate mouth operations, all pipetting must be done using automatic pipettes. Most of those shown in Fig. 4 will be familiar to those who work in a bacteriological laboratory.

Any laboratory in which radioisotopes are used must possess a contamination monitor. The exception is in the case of work involving only the radioactive isotope of hydrogen, tritium. The *beta* energy of this isotope is so low that normal monitoring equipment will fail to detect it and the only way in which monitoring may be carried out is by making periodic

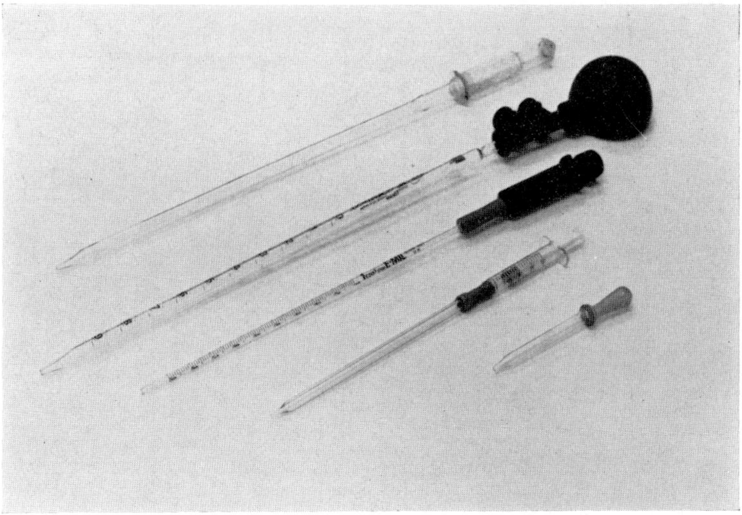

FIG. 4. Selection of pipettes used in the radiochemical laboratory.

checks of tritium activity in the urine of personnel. Such measurements may be made with a liquid scintillation counter. Work with all other isotopes requires a monitor which generally consists of either a Geiger-Müller counter or a scintillation counter attached to a rate-meter. The instrument should preferably have an audible output and typical examples are shown in Fig. 1. These instruments can readily detect small areas of contamination on benches and apparatus or on protective clothing and hands.

Decontamination of apparatus and personal decontamination

The maximum permissible levels of contamination for apparatus and the skin are given in Table 5. This table also gives the count rates to be expected with the Geiger-Müller and scintillation counter contamination monitors of Fig. 1, due to these levels of contamination, After use, all apparatus and the person, especially the hands, should be checked for contamination and if it is present above permissible levels efforts should be made to remove it.

Most radioactive contamination is easily removed from apparatus using

TABLE 5. Maximum permissible levels of contamination for apparatus and skin

Area	Isotope	Maximum Permissible Level (μCi/cm^2)	Approximate count rates (counts per minute) using: G-M tube (halogen quenched) 3·2 cm^2 area window of thickness 4·5 mg/cm^2	Approximate count rates (counts per minute) using: ZnS/plastic phosphor scintillation counter 49 cm^2 area covered with mylar film 2 mg/cm^2
Inactive area (equipment)	Phosphorous-32 (energetic β^-)	10^{-4}	370	6200
	Carbon-14 (weak β^-)		20–24	insensitive
	Sodium-22 (β^+/γ)		240	1450
Active area (equipment)	Phosphorus-32	10^{-3}	3700	62,000
	Carbon-14		240	insensitive
	Sodium-22		2400	14,500
Hands	Phosphorus-32	3×10^{-2} μCi/hand or approx. 10^{-4} μCi/cm^2	370	6200
	Carbon-14		20–24	insensitive
	Sodium-22		240	1450

detergents, especially the surface active types such as Decon-90 (Medical Pharmaceutical Developments Ltd., Portslade, Brighton, England). Some materials, such as phosphorus-32 labelled phosphate, can be strongly adsorbed on glassware but can generally be removed with concentrated acids. Concentrated hydrochloric acid will remove phosphate but more stubborn contamination can be removed by soaking the apparatus in chromic acid solution. Apparatus not resistant to acids can be decontaminated by soaking in a solution of an inactive carrier for the contaminating isotope so that chemical exchange can take place. Chelating agents such as a 5% (w/v) solution of ethylene diamine tetra-acetic acid (EDTA) are also useful. Highly contaminated apparatus, which such procedures have failed to decontaminate successfully, may either be stored to permit radioactive decay or discarded as radioactive waste.

With proper techniques radioactive contamination of the person is rare but if it does occur it must if possible be removed. Contamination of the skin should be treated by washing with soap, then detergent and then, if necessary, by gentle scrubbing with a soft nail brush. If repetition of these treatments fails to remove the activity, then the area may be treated with a potassium permanganate solution followed by de-colorization with sodium metabisulphite solution. This treatment should be confined to the hands and used only as a last resort. Decontamination procedures should never be carried on to such an extent that skin damage is caused. It is very unlikely that significant activity will remain after such treatments but, if it does, it is unlikely to be an ingestion hazard because of the difficulty in removing it. The natural sloughing off of dead skin cells will effect its removal within a few days. This process may be aided by covering the area, if small enough, with a zinc oxide plaster.

Code of Practice for Persons Exposed to Ionizing Radiations

All persons working with radioisotopes in research or teaching establishments should conform to the provisions of the "Code of Practice for the Protection of Persons exposed to Ionizing Radiations in Research and Teaching" (Anon, 1964). This requires that persons likely to receive radiation doses at the level appropriate for occupational radiation workers should be "designated", that they should be subjected to personal monitoring and that radiation dose records for them should be kept by their employer. Whether or not a person working with radioactive materials should be "designated" depends upon the radiation levels due to the radioactive materials at his place of work. It is generally unlikely that tracer work in a bacteriology laboratory would require "designation" or personal monitoring unless a large volume of work was being carried out or the radioactive materials involved were of high toxicity. The safety provisions of the Code, however, must still be followed and the outline given above is based upon its recommendations.

Personal dosimetry for persons working with energetic β-particle or β, γ-emitters can readily be carried out by requiring the worker to wear a film badge such as that shown in Fig. 5. This consists of a small piece of

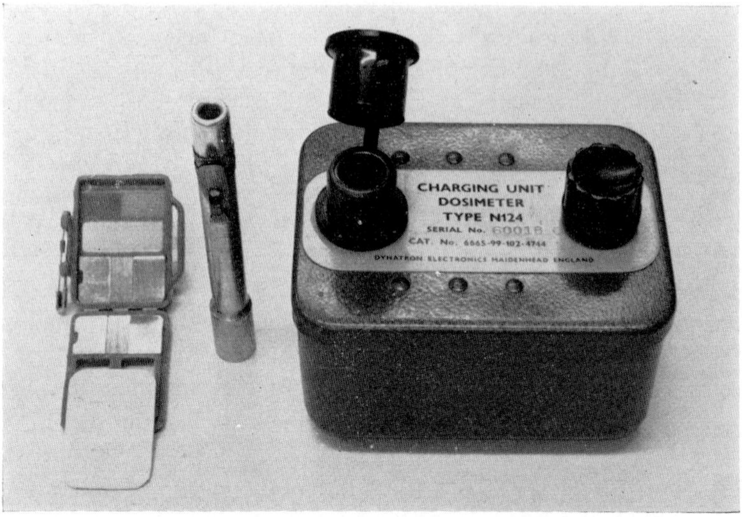

FIG. 5. Personal monitoring equipment. A film badge and holder (left), pocket dosimeter (centre), and a dosimeter charging unit (right).

X-ray film encased in a special nylon holder which contains filters to enable the interactions of γ-rays and β-particles with the film to be determined. The effect of these interactions is to produce a latent image in the film so that blackened regions appear upon development. The dose which the wearer has received can be determined from the intensity of the blackening. In the U.K., the National Radiological Protection Board will supply, process and report upon such films used for radiation dose measurement. Personal dosimeters of the pen type (also shown in Fig. 5) depend for their operation on the discharge by the ionizing radiation of a small quartz fibre electroscope. They are sensitive only to γ-rays and need only be used when very high activities of γ-ray emitters are used.

The film badge is not sensitive to low energy β-particles and personal dosimetry for low energy β-emitters such as tritium, carbon-14 and sulphur-35 can only be carried out by determining the ingested activity by urine analysis. The levels of activity at which such analysis becomes advisable are given in Table 6.

TABLE 6. Levels of radioactivity above which various safeguards are required*

Radionuclide	Fume cupboard required	Glove box required	Personnel monitoring and/or shielding against external γ-radiation	Occasional excretion radio assay spot checks of personnel	Routine excretion assay of all personnel
^{3}H	1 Ci	10 Ci	—	10 Ci	10^{2}Ci
^{14}C	1 Ci	10 Ci	—	10 Ci	10^{2}Ci
^{35}S	10^{-2}Ci	0·1 Ci	—	0·1 Ci	1 Ci
^{65}Zn	10^{-2}Ci	0·1 Ci	—	↓	↓
^{59}Fe	10^{-2}Ci	0·1 Ci	0·2 Ci	↓	↓
^{32}P	10^{-3}Ci	10^{-2}Ci	10 Ci	10^{-2}Ci	0·1 Ci
^{36}Cl	↓	↓	—	↓	↓
^{45}Ca	↓	↓	10 Ci	↓	↓
^{137}Cs	↓	↓	—	↓	↓
^{60}Co	↓	↓	0·3 Ci	↓	↓
^{131}I	↓	↓	0·4 Ci	↓	↓
^{90}Sr	10^{-5}Ci	10^{-4}Ci	10 Ci	10^{-4}Ci	10^{-3}Ci

* Data from Brodsky (1965).

The hazards due to labelled compounds

This discussion has dealt mainly with the hazards due to radioactive nuclides in the elemental form. Complex organic materials such as carbon-14 or tritium-labelled pyrimidine bases and their derivatives are being increasingly used in biochemistry and are commonly to be found in bacteriological laboratories. If these compounds are ingested, they become incorporated into nucleic acids, generally of actively dividing tissues. The labelled atom in simple inorganic form undoubtedly would not generally gain access to such vulnerable and genetically significant molecules. The hazard in using such materials must therefore be regarded as being considerably greater than for simple labelled forms and consequently extra care must be taken in dealing with them. As yet no definite guidance has been issued on the relative hazards of these biochemically significant labelled materials.

Registration of Premises for the Use of Radioactive Materials and Authorizations for Radioactive Waste Disposal

In the U.K., the Radioactive Substances Act (1960) requires that premises in which radioactive materials are used should be registered with the Department of the Environment. The Radiochemical Inspectorate of this Department is also responsible for the authorization of radioactive waste disposal from such premises. Since specific authorization is required for all radioactive waste disposal from registered premises, it is not possible to give specific guidance on waste disposal. The following must be considered a general guide-line only.

In all work with radioactive materials some radioactive waste must be created although, in the case of tracer laboratories, the amounts are generally small. The waste will arise in the form of liquid (suspension or solution), solid and gaseous wastes. Since comparatively large activities of liquid waste can generally be released into a laboratory drainage system, liquid waste is the simplest to dispose of and waste should, if possible, be kept in this form. Actual permissible levels will be dependent on the establishment but disposal of up to several mCi/month from a tracer laboratory would not be unusual. Only small amounts of radioactive solid waste may be released to a Local Authority's waste collection service. The waste must be mixed with inactive solid wastes and the maximum activities that may be disposed of by this means are limited to a few μCi of activity in a standard dust-bin (3 ft^3). Special provision for the controlled burial of larger amounts of activity must be made with the Radiochemical Inspectorate. Incineration of solid wastes with a very efficient incinerator may be acceptable but radioactive ash may be produced which would present a further disposal problem and a contamination hazard. Contaminated animal carcasses can most conveniently be disposed of by maceration in a waste disposal unit provided that the use of such units is approved by the Local Authority. The release of gaseous radioactive wastes into the atmosphere depends upon the situation of the laboratory with respect to surrounding buildings but generally the permissible levels of activity are very low. Waste disposal does not generally present great problems for the tracer laboratory but in all cases the limits and methods of disposal must be agreed with the Radiochemical Inspectorate of the Department of the Environment.

References

ANON (1959). *Recommendations of the International Commission on Radiological Protection.* (ICRP report of Committee II). London: Pergamon.

ANON (1964). *Code of Practice for the protection of persons exposed to Ionising*

Radiations in Research and Teaching. Department of Employment and Productivity. London: HMSO

Anon (1966). *Recommendations of the International Commission on Radiological Protection*. (ICRP publication 9). London: Pergamon.

Bowen, H. J. M. & Gibbons, D. (1963). *Radioactivation Analysis*. Oxford: Oxford University Press.

Brodsky, A. (1965). Determining industrial hygiene requirements for installations using radioactive material. *Am. ind. Hyg. Ass. J.*, **26**, 294.

Gorsuch, T. T. (1968*a*). *Radioactive tracers in chemical analysis*. RCC Review 5, Amersham: The Radiochemical Centre.

Gorsuch, T. T. (1968*b*). *Radioactive isotope dilution analysis*. RCC Review 5, Amersham: The Radiochemical Centre.

Hevesy, G. (1962). *Adventures in Radioisotope Research*. (Collected Papers), London: Pergamon.

Oldham, K. G. (1968). *Radiochemical methods of enzyme assay*. RCC Review 5, Amersham: The Radiochemical Centre.

Sherwood, R. J. (1959). *A short course in Radiological Protection*. Harwell: A.E.R.E.

Wilson, B. S. (ed.) (1966). *Radiochemical Manual*. 2nd Ed. Amersham: The Radio-chemical Centre.

Relevant Acts and Regulations

Radioactive Substances Act, 1960. HMSO, London.

The Ionising Radiations (sealed sources) Regulations, 1961. (S.I. 1961 No. 1470) HMSO, London.

The Unsealed Radioactive Substances Regulations, 1965. HMSO, London.

Preservation of Fungal Cultures and the Control of Mycophagous Mites

R. C. CODNER

Microbiology Section, School of Biological Sciences, University of Bath, Bath BA2 7AY, Somerset, England

In the demonstration originally presented under this title two minor aspects of fungal culture maintenance were illustrated namely the use of sterile soil for the preservation of mould cultures and certain precautions which may be used to combat the invasion of fungus culture collections by mycophagous mites. The Editors of the present volume have indicated that contributors need not limit themselves to the material presented in the demonstration and therefore a more general consideration of the methods used in the preservation of living fungal culture collections is given here and the discussion of the problem of mite infestation is somewhat extended.

Preservation

A number of previous publications dealing with fungus culture collections contain much helpful information, thus Clark and Loegering (1967) survey the functions and organization of the American Type Culture Collection and give an account of the methods of preservation in use in this collection. This publication also includes a useful list of the world's major fungal culture collections with their addresses. A more general view of the problems facing those responsible for culture collections may be found in the account (Martin, 1962) of the Ottawa Conference on culture collections. Some of the methods used at the Commonwealth Mycological Institute culture collection are discussed by Dade (1960) and other useful accounts of the preservation of fungal cultures and the maintenance of fungal culture collections are given by Fennell (1960) and Calam (1965). The latter discusses in particular the problems of maintenance of high yielding strains of fungi used in the production of commercially important metabolites. It is evident that a diversity of methods are currently used for the preservation of fungal cultures. These methods fall roughly into the following categories.

Storage of cultures on agar slants

In this method of culture storage, no special steps are taken to extend the useful life of individual sub-cultures other than the selection of suitable media such as Potato-Carrot Agar (Dade, 1960). Potato (20 g) and Carrot (20 g) are cleaned and boiled in 1 litre of water, filtered and solidified with agar (2% w/v). This medium discourages luxurious growth of aerial hyphae without depressing the sporulation of fungi grown upon it. Well grown cultures are stored either at room temperature or 4°. Cultures on agar slopes in test tubes should have the plugs and necks of the tubes wrapped in grease-proof paper to reduce accumulation of dust and to reduce moisture loss. Cultures on slopes in screw capped bottles should not have the caps screwed tight during storage.

The method has the advantages that no special equipment is needed, that it is simple to operate and that cultures are at all times available for immediate use or subculture. The disadvantages of the method include the relatively limited period for which cultures of many fungi remain viable, the tendency of high yielding strains of industrially important fungi to lose their productive capacity on repeated subculture and the liability of cultures particularly when stored at ambient temperature to be attacked by mites. It is suggested by Dade (1960) as a rough guide that cultures should be transferred to fresh medium at intervals of about 6 months.

Storage of agar slant cultures under mineral oil

The useful storage life of agar slant cultures of most fungi so far tested is greatly extended by the simple process of covering well grown slant cultures with mineral oil (medicinal liquid paraffin) so that the top of the agar slant is not less than 0·5 cm below the surface of the oil. The liquid paraffin requires careful sterilization by prolonged autoclaving for 2 h/121°. It has been found useful to add a small quantity of water to each container before sterilization thus ensuring that the oil is subject to moist heat during sterilization. Fennell (1960) suggests that the oil should be dried after autoclaving by heating to 170° for 1–2 h in an oven. The need for this drying seems doubtful since the slope to which the oil is to be added will already contain some 95% water.

When treating a number of cultures it is advisable to sterilize the oil in suitable amounts in separate small containers for use in covering individual cultures to prevent cross contamination. Dade (1960) suggests that when using 1 oz Universal bottles for oil covered slopes it is advisable to use caps from which the rubber wads have been removed to prevent the oil becoming contaminated by materials dissolved from the rubber. Cultures under oil may be stored at room temperature though for most fungi and yeasts storage at 4° would increase the period of survival. The value of storage under oil is attributable first to the prevention of drying of the agar and secondly to the reduction of metabolic activity of the culture due to the restriction of the oxygen available to the culture.

Cultures under oil may be used for the inoculation of further sub-cultures at any time and no special equipment is required to transfer from such cultures. It will however be noted that a small drop of oil is inevitably to be found at the point of inoculation of cultures prepared from slants stored under oil. The inoculum however soon grows away from the oily area and normal growth occurs. When flaming the loop after transfer from oil covered cultures great care should be taken to prevent contamination of the work bench and the operator by spattering of the residual oil droplets on the loop—for further discussion of this point, see p. 13. Cultures stored under oil are safe from the depredations of mites which are neither attracted to nor can penetrate the mineral oil. It is not easy to generalize on the survival period of cultures stored by this method though reports indicate that a high proportion of fungi may be expected to survive for periods of up to 10 years. The method is of special value for maintaining cultures which remain purely mycelial (as do many *Basidiomycetes*) and which are not conveniently freeze dried. Reischer (1949) has advocated the use of oil for the preservation of members of the *Saprolegniaceae*, a group which requires constant attention when maintained by serial transfer.

Storage of cultures grown and dried on sterile soil

Cultures of a number of fungal genera have been preserved successfully as soil cultures using the method originally described by Greene and Fred (1934). The method of preparation of soil cultures involves sieving an ordinary loamy garden soil in the moist state through a 2 mm mesh sieve. The sieved soil is then air dried at room temperature. The air dried soil is filled into test tubes, 1 oz screw capped bottles or other suitable containers. An approximately measured quantity of soil (not more than three-quarters of the volume of the container) is added to each container and sufficient distilled water is then added to bring the moisture content of the soil to 20%. The containers need shaking to distribute the water evenly throughout the bulk of the soil. The containers, closed with caps or cotton wool plugs, are then sterilized by autoclaving at 121°/1 h on three successive days. Sample containers taken at random from the batch are tested for sterility by the aseptic addition of Nutrient Broth until the soil is covered with a layer of broth in which growth due to bacterial contamination can be detected by the formation of turbidity or a pellicle after incubation. Since the most likely organisms to survive this exaggerated Tyndallization process are the more heat resistant bacterial spores it can be assumed that, if bacterial growth is not detected in any sample, the less heat resistant bacteria and fungi will also be destroyed. The heat sterilization of the soil also overcomes the effect of soil fungistasis (Dobbs and Hinson, 1953) which would otherwise inhibit the germination of spores added to the soil.

The moist sterile soil may be inoculated with a spore suspension using up to 1 ml/5 g of sterile soil. Soil cultures may also be inoculated with dry spores by loop but growth is likely to be established more slowly. The inoculated soil is then incubated and allowed to dry out slowly at room temperature. During incubation some fungi produce considerable growth visible in the interstitial spaces between the soil crumbs. Fungi not producing visible growth will nevertheless produce an adequate crop of spores to ensure satisfactory storage.

After incubation the soil cultures are best stored in a dry state at 4° though cultures will survive quite satisfactorily at room temperature. Many fungi and actinomycetes stored by the soil culture method, show much less variation in their morphological characters and also in their biochemical characters such as antibiotic production than they show when maintained on agar slants with repeated transfer. There are many reports of soil cultures of fungi and actinomycetes surviving for several years without the necessity for subculturing and the writer has recovered a strain of *Cephalosporium* from a 17 year old soil culture.

Soil cultures provide a convenient source of inoculum for the preparation of the many identical cultures required over long periods of time in the industrial production of antibiotics and other microbial metabolites. A few grains of soil from a stock soil culture provide an adequate inoculum for a single agar slope culture and many hundreds of such identical subcultures can be prepared from a single soil culture, in a 1 oz bottle, over an extended period. The grains of soil forming the inoculum can be transferred by a sterile loop moistened with sterile distilled water or with condensate from the agar to be inoculated.

Soil cultures have the disadvantage that it is not possible to detect contamination or to check the colonial characters of the culture visually without resort to plating out. Until thoroughly dried, soil cultures are not immune to attack by mites.

In another, rather similar method of culture preservation a concentrated suspension of spores of the organism to be preserved is added to sterile soil which is immediately dried out, in a desiccator or by freeze drying, without opportunity for the spores to germinate. Other materials have been compared with soil as a support medium in dried spore preparations of this kind but have proved less satisfactory. The particular virtue of soil in the preservation of cultures may lie in the peculiar adsorption and ion exchange properties of the clay colloids which may remove and bind autotoxic metabolites and "staleing products".

Culture preservation by drying other than by drying from the frozen state

It is clear from many observations that agar cultures of fungi can survive for periods of up to several years on drying out through neglect (Fennell, 1960). It is however unwise to rely on uncontrolled natural desiccation as a means of preserving fungal cultures and a number of methods of accelerated drying using desiccants either with or without the application of reduced pressure have been used in the preservation of cultures of microorganisms.

Perkins (1962) gives details of one such method for the preservation of stock cultures of *Neurospora* which is applicable to both conidiating strains and to mycelial mutants. The method, which is the essence of simplicity, involves adding a suspension of spores or mycelial fragments to silica gel which has been dry sterilized at 180° for $1\frac{1}{2}$ h. The gel granules (6–12 mesh; without indicator dye) are held in plugged test tubes not more than half filled. The use of non fat milk in the suspending medium is recommended to provide protective colloids. The suspension is first prepared in distilled water so that it can be seen to be evenly distributed and of a suitable density. An equal volume of sterile milk is then mixed with the suspension which is then added to the silica gel granules at a rate of *c.* 1 ml of suspension to 4 g of dry silica gel. The silica gel granules are arranged as a slope, by tapping the tube in a sloped position, so that the suspension may be distributed evenly by dropwise addition from a pipette over the length of the slope of granules. This minimizes heating due to the hydration of the silica gel. The tubes are held for 1 week at room temperature before being checked for viability. The plugged mouths of the tubes are then sealed with Parafilm (available from Gallenkamp Ltd., Christopher Street, London E.C.2). Over 80% of cultures of *Neurospora* examined by Perkins remained viable for up to 6 years by this method.

Dr. C. F. Roberts of the Department of Genetics, University of Leicester, has prepared a useful note communicated privately in which this technique is modified using materials of British origin and specification. Thus nonfat milk is prepared using a 5% (w/v) solution of Cadbury's Marvel milk powder, while silica gel (purified; 6–22 mesh; without indicator; Hopkin and Williams, Freshwater Road, Chadwell Heath, Essex) is suitable for this technique. To hold 4g amounts of silica gel, 2 dram vials (Series SNB 17 × 58 mm C94 Spec. No. 6/H/0001 closure 3G/H/2053/0; Johnsen and Jorgensen Ltd., Herringham Road, London S.E.7) are recommended which have cap liners covered by metallic foil. Dr. Roberts also suggests that heating problems may be reduced by holding the vials of silica gel in an ice bath for 30 min before adding the fungal suspension and returning the vials to the ice bath for 15 min immediately after the addition of the suspension. Roberts recommends that the dried cultures should be stored in airtight boxes containing silica gel with indicator at 4°. This modified technique has been found valuable by Roberts in maintaining the viability and cultural characters of several hundred strains of *Aspergillus nidulans*.

In certain early methods of culture preservation by vacuum desiccation (Stamp, 1947; Rhodes, 1950*a*, *b*) it is difficult to decide whether the sample does in fact freeze by evaporative cooling during the drying process and thus whether drying takes place from the frozen state as in the methods discussed in the next section.

Preservation of fungi by freeze-drying

The technique of freeze-drying, well known to those concerned with the preservation of bacterial cultures, has been widely adopted for the preservation of fungal cultures, particularly high yielding strains used in industrial processes. It is not intended here to provide a detailed discussion of the factors involved in the survival of fungi subjected to freeze drying as the review by Mazur (1968) already treats this problem most adequately. From a practical standpoint the choice of technique for freeze drying fungi is often dictated by the equipment available either commercially or within a particular laboratory. Techniques fall broadly into two categories; those involving pre-freezing of the suspension of fungal cells either as a plug, a wedge or a shell prior to transfer to the vacuum system and those techniques in which freezing takes place as a result of evaporative cooling as the system is evacuated. If serious foaming of the sample is to be avoided the sample must, in the evaporative freezing technique, be subjected to centrifugal force to break the foam (Greaves, 1944). When using evaporative cooling the ultimate low temperature reached by the sample is dependent on the freezing point of the sample and is unlikely to reach such low temperatures as are reached when using pre-freezing methods which usually employ as the freezing agent solid carbon dioxide with an organic solvent such as methyl cellosolve, e.g. the method of Wickerham and Andreasen (1942). In the techniques involving evaporative cooling the vacuum system must be adequate to remove rapidly some 20% of the water in the sample before freezing can take place otherwise cells may be exposed to high concentrations of solutes in the liquid phase for long periods before freezing occurs (Barratt and Beckett, 1951).

The medium in which organisms are freeze-dried may have a considerable effect on the proportion of cells which survive both initial drying and subsequent periods of storage. It would appear either that fungi have less fastidious requirements in this respect than do bacteria or that many investigations on the survival of fungi have been based on the less rigorous criterion of ability to recover viable cells after storage and not on the proportion of the original population surviving. Satisfactory preservation of many strains of *Penicillium* and *Aspergillus* has been reported by Mehrotra and Hesseltine (1958) using undiluted freshly prepared beef serum. The use of horse or human serum has also been shown to be satisfactory by the author (unpublished) for the routine preservation of many *Fungi Imperfecti* including particularly members of the form-genus *Cephalosporium*. Satisfactory results have also been obtained by freeze-drying fungi suspended in "*Mist. desiccans*", suggested by Fry (1951), which consists of one part Nutrient Broth and three parts serum; the mixture containing overall 7·5% (w/v) glucose. This is conveniently prepared using a 30% (w/v) solution of glucose in Nutrient Broth, one part of this solution being added to three parts of serum. The mixture is sterilized by filtration and stored at −20°C.

Other materials suggested as being of value for use in suspending media for freeze-drying fungi include dextrin, alginate, peptone, glucose and sucrose. The value of strongly reducing substances such as ascorbic acid and thiourea suggested for use in the freeze-drying of bacteria (Naylor and Smith, 1946; Stamp, 1947) has not been proved with fungi. It may be speculated that these substances are of greater value where the dried culture is sealed down under air rather than in high vacuum.

In view of the very high initial survival rates of spores of *Aspergillus flavus* freeze-dried in the absence of any aqueous suspending medium, reported by Mazur (1968), it would seem desirable to examine further the freeze-drying of other fungal spores dried in the absence of any suspending medium as a method for routine preservation. Maximal survival of freeze-dried cultures is obtained when the ampoules are sealed under high vacuum rather than in air or an inert gas such as nitrogen at atmospheric pressure. Individual ampoules may be tested for satisfactory vacuum using a high frequency electric discharge tester. Evidence on the effect of storage conditions on the survival of freeze-dried fungal cultures is limited but on theoretical grounds it would seem desirable to store cultures in the dark and at refrigerator temperature of 4°.

Preservation by freezing and storage at very low temperatures

With the increased availability of liquid nitrogen and the commercial availability of suitable apparatus for the storage of sealed ampoules at the boiling point (−196°) of liquid nitrogen, a further possible method for the preservation of microbial cultures has become available. The method should be applicable to any organism which will survive freeze-drying and is applicable to certain fungi which do not sporulate readily in culture, such as *Rhizoctonia* and many other basidiomycetes and to fungi which normally do not freeze-dry satisfactorily such as *Pythium* and *Phytophthora* spp. Clark and Loegering (1967) cite reports of satisfactory storage of a number of fungi by this method for periods of up to 5 years.

The apparatus for the application of this technique of storage consists of specially designed and insulated vacuum flasks in which large numbers of small ampoules can be stored in lift-out canisters. The equipment requires no maintenance other than weekly topping with liquid nitrogen and in emergency this topping-up may be delayed for considerably longer periods depending on the size of the flask. The vacuum flasks should also be emptied annually for inspection and should be flushed out with dry gaseous nitrogen before refilling.

The main factors influencing the survival of fungi after freezing to low temperature and thawing, appear to be the rate at which both freezing and thawing take place and the protective medium chosen for suspending the fungus. Thus for *Aspergillus flavus* spores and *Saccharomyces cerevisiae* cells, Mazur (1960) points out that relatively slow cooling (1°/min or less) gave far higher survival than did rapid cooling at 300°/min, while rapid warming (at 1100°/min) of frozen cultures gave higher survival than did slow warming at 1°/min. Controlled slow cooling may be achieved with relatively simple apparatus such as that described by Greaves, Nagington and Kellaway (1963). Results of preservation of cultures by storage under liquid nitrogen and technical details of one method of processing cultures is given by Hwang (1966).

Both the freeze-drying technique and the use of liquid nitrogen refrigeration for the storage of cultures have the advantages of very much extended storage life, limitation of mutation of cultures and, due to storage in sealed ampoules, freedom from contamination and from attack by mites. Both techniques require some specialized equipment and the low temperature storage method has a continuing requirement for a supply of liquid nitrogen.

From the above discussion it will be clear that there is no one universally applicable simple method of culture preservation for use where a wide range of fungal genera is to be maintained. However by a judicious choice from the above methods the labour and anxiety involved in the maintenance of fungal culture collections may be much reduced.

The Control of Mycophagous Mites

Whenever mycologists discuss the generalities of laboratory housekeeping peculiar to their branch of microbiology the topic of infection of cultures by mites is almost certain to be raised. Mites are to quote Dade (1960) "the hated enemies of all mycologists. They are attracted by the odours of fungi and enter culture vessels bringing on their legs contaminating bacteria and moulds."

FIG. 1. Adult mites are from 250–500μ in length.

Acarine mites have, in the adult stage, four pairs of legs though in an earlier stage of development only three pairs of legs are present. Adult mites (Fig. 1) are *c.* 250–500μ long and are thus only just visible to the naked eye and may pass unnoticed in the early stages of an infestation unless cultures are carefully scrutinized, preferably with a hand lens. Often the first indication that all is not well is the contamination of cultures with other microorganisms. Where contaminated cultures consist of isolated colonies the contamination often shows up as irregular tracks of colonies arising from the movement of the contaminated mites over the agar surface. While mites are usually attracted to established fungal cultures, on which

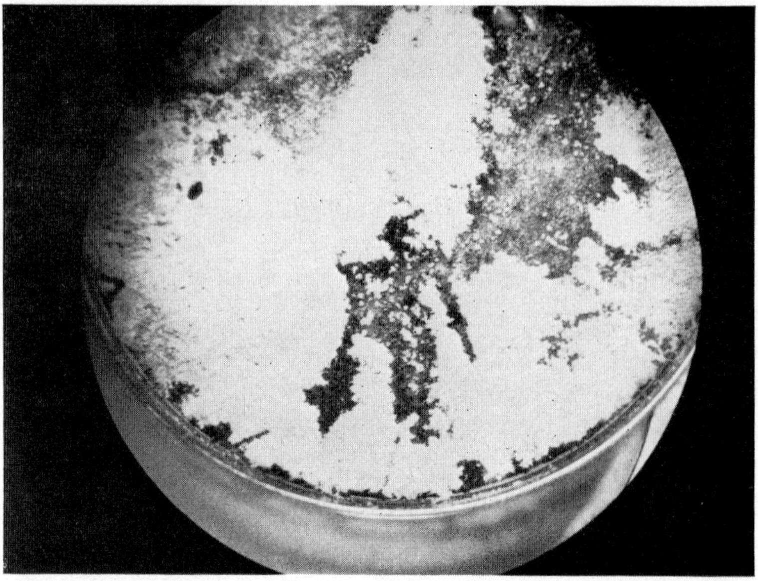

FIG. 2. Result of mite infestation. Irregular dark area resulting from "browsing" by mites in a Petri dish culture of *Penicillium*.

they may produce visible bare patches by browsing (Fig. 2), the writer has also observed mite trails on uninoculated agar in cotton wool plugged vessels. The cotton wool plug does not provide an adequate barrier to penetration by mites.

Mites are widespread in nature and are globally distributed. Many thousands of species, fortunately not all associated with fungi, have been described. Members of the genera *Tarsonemus*, *Glyciphagus* and *Histiostoma* are particularly implicated as mycophagous forms. Most decaying vegetation, mouldy food materials, mouldy cereal grains, hard cheese and dried meat products such as salami may carry mite infestations. Samples of this

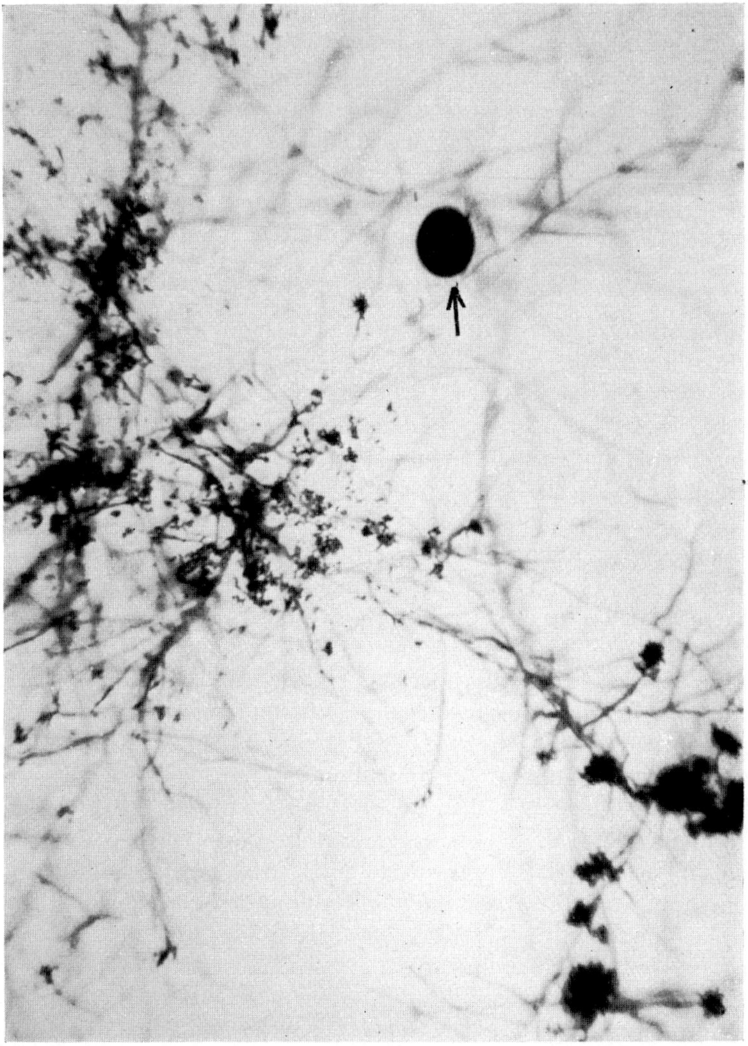

FIG. 3. Mite egg. Arrow indicates mite egg among hyphae of a fungal culture.

kind brought into the laboratory may act as a source of general infestation unless treated with extreme care. The prolonged incubation, seldom less than 7 days, required for the development of fungal cultures also enables the eggs (Fig. 3) of the first invading mites to hatch out and become mobile and move from one culture to another or into crevices in the incubator.

The first precaution that can be taken is the maintenance of absolute cleanliness of working surfaces. This, however, is no deterrent to the movement of mites and it is useful to wipe over working surfaces and woodwork with Tractor Vapourizing Oil (TVO). This material has a strong though not unpleasant smell and is strongly repellant to mites. In addition to the treatment of working surfaces and incubator shelves with TVO, Dade (1960) recommends the use of small bottles of TVO equipped with a short length of lamp wick (a twist of cotton wool, cloth or pipe cleaners will serve almost as well as a wick) placed in incubators to maintain a low concentration of vapour. Curl (1958) recommends the use of a 3–3·5% solution of Kelthane E 1, 1 *bis* (chlorophenyl)—2,2,2 trichloroethanol (Röhm and Haas) for application to shelves and tables by spraying once a fortnight and for spraying incubators and transfer rooms at weekly intervals. Kelthane preparations for the control of red spider mite in horticultural crops are available from the Murphy Chemical Company, Wheathampstead, St. Albans, or from Pan Britannica Industries Ltd., Britannica House, Waltham Cross, Herts.

Mites may be prevented from entering incubators from floors by standing the incubator on bungs or blocks, in turn standing in dishes or trays of lubricating oil or medicinal paraffin to form a moat (Fig. 4). A similar device may be used to isolate infested samples on a tray either on the bench or during incubation on an incubator shelf. Snyder and Hansen (1946) describe the use of cigarette paper barriers applied to the mouth of plugged test tubes. After the plug has been pushed into the tube and flamed off level with the mouth of the tube, the hot rim of the tube is pressed into a 20% (w/v) gelatin gel in water containing 2% (w/v) copper sulphate. The coated rim of the tube is then pressed on to a square of white cigarette paper to form a seal. Excess paper may be burned off to leave a neat circular mite-proof seal. Raper and Fennell (1965) recommend the poisoning of cotton wool plugs with the following mixture before adding cultures to a collection: mercuric chloride, 0·5 g; glycerine, 5 ml, and 95% alcohol, 95 ml coloured with an aniline dye.

The problem of eradicating mites from infested cultures has been considered by a number of workers. Some success has been obtained by sealing infested cultures in a large tin with a small quantity of *p*-dichloro benzene (PDCB) crystals overnight (Smith, 1960). A similar method described by Benedek (1963) uses thymol crystals for the fumigation of infested cultures. Both of these substances have some toxicity to fungi and PDCB is considered to cause mutation. Benedek, however, states that the exposure of cultures to thymol vapour had no effect on the growth characteristics of the dermatophyte fungi used in his experiments. After ridding cultures of the infesting mites the cultures must immediately be plated out to eliminate contaminating organisms introduced by the mites.

FIG. 4. Corner of incubator supported on a bung in a moat of oil to prevent mite invasion.

Many other substances have been suggested or tried for the elimination or control of mites including hexachlorocyclohexane, dichloroethane, methyl bromide, xylene and benzene hexachloride (Gammexane). Rhodes and Fletcher (1966) recommend the use of suitable agar culture media containing 0·025% (w/v) Kelthane (based on the active ingredient) for the elimination of mites from infested cultures. Two or more sub-cultures on media containing Kelthane may be necessary for complete disinfestation.

In view of the number of acaricidal preparations available for horticultural use some of these might well be of value in the control of mycophagous mites.

It is however preferable to avoid mite infestation by constant vigilance and a high standard of laboratory housekeeping together with the use of the mite repellants and control practices indicated, rather than face the problem of disinfestation and subsequent purification of a large number of cultures. It also leads to peace of mind if at least some cultures of each organism are preserved by freeze drying or stored under liquid nitrogen so as to be inaccessible to invasion by mites.

References

Barratt, A. S. D. & Beckett, L. G. (1951). Aspects of the design of freeze drying apparatus. In *Freezing and Drying* (R. J. C. Harris, ed.) pp. 41–49. London: The Institute of Biology.

Benedek, T. (1963). Use of thymol as an acaricidal agent against infestation of fungus cultures and mycotheca with Acari (Mites) *Mycopath. Mycol. appl.* **19,** 87.

Calam, C. T. (1965). The selection, improvement and preservation of micro-organisms. In *Progress in Industrial Microbiology* (D. J. D. Hockenhull, ed.) Vol. 5. London: Iliffe.

Clark, W. A. & Loegering, W. Q. (1967). Functions and maintenance of a type culture collection. *A. Rev. Phytopath.*, **5,** 319.

Curl, E. A. (1958). Chemical exclusion of mites from fungal laboratory cultures *Pl. Dis. Reptr.*, **42,** 1026.

Dade, H. A. (1960). Laboratory methods in use in the culture collection C.M.I. In *Herb I.M.I. Handbook*. Kew, Surrey: Commonwealth Mycological Institute.

Dobbs, C. G. & Hinson, W. H. (1953). A widespread fungistasis in soils. *Nature, Lond.*, **172,** 197.

Fennell, D. I. (1960). Conservation of fungus cultures. *Bot. Rev.*, **26,** 79.

Fry, R. M. (1951). The influence of suspending medium on the survival of bacteria after freezing and drying. In *Freezing and Drying* (R. J. C. Harris, ed.) pp. 107–115. London: The Institute of Biology.

Greaves, R. I. N. (1944). Centrifugal vacuum freezing. Its application to the drying of biological materials. *Nature, Lond.*, **153,** 485.

Greaves, R. I. N., Nagington, J. & Kellaway, T. D. (1963). Preservation of living cells by freezing and by drying. *Fedn Proc. Fedn Am. Socs exp. Biol.*, **22,** 90.

Greene, H. C. & Fred, E. B. (1934). Maintenance of vigorous mould stocks. *Ind. Engng. Chem., ind. Edn.*, **26,** 1297.

Hwang, S. W. (1966). Long term preservation of fungus cultures with liquid nitrogen refrigeration *Appl. Microbiol.*, **14,** 784.

Martin, S. M. (ed.) (1962). *Culture collections: perspectives and problems*. Proceedings of the specialized conference on culture collections Ottawa. Toronto, Canada: Univ. Toronto Press.

Mazur, P. (1960). Physical factors implicated in the death of microorganisms at sub-zero temperatures. *Ann. N.Y. Acad. Sci.*, **85,** 610.

Mazur, P. (1968). Survival of fungi after freezing and desiccation. In *The Fungi*

(G. C. Ainsworth and A. S. Sussman, eds) Vol. 3. Ch. 14, pp. 235–394. New York and London: Academic Press.

MEHROTRA, B. S. & HESSELTINE, C. W. (1958). Further evaluation of the lyophil process for the preservation of aspergilli and penicillia. *Appl. Microbiol*, **6,** 179.

NAYLOR, H. B. & SMITH, P. A. (1946). Factors affecting the viability of *Serratia marcescens* during dehydration and storage. *J. Bact.*, **52,** 565.

PERKINS, D. D. (1962). Preservation of *Neurospora* stock cultures with silica gel. *Canad. J. Microbiol.*, **8,** 591.

RAPER, K. B. & FENNELL, D. I. (1965). *The Genus* ***Aspergillus*** pp. 65–66. Baltimore: Williams and Wilkins.

REISCHER, H. S. (1949). Preservation of *Saprolegniaceae* by the mineral oil method *Mycologia*, **41,** 177.

RHODES, A. & FLETCHER, D. L. (1966). *Principles of Industrial Microbiology* p. 38. Oxford: Pergamon Press.

RHODES, M. (1950*a*). Preservation of yeasts and fungi by desiccation. *Trans. Br. mycol. Soc.* Sac., **33,** 35.

RHODES, M. (1950*b*). Viability of dried bacterial cultures. *J. gen. Microbiol.*, **4,** 450.

SMITH, G. (1960). *Industrial Mycology* p. 307. London: Arnold.

SNYDER, W. C. & HANSEN, H. N. (1946). Control of culture mites by cigarette paper barriers. *Mycologia*, **38,** 455.

STAMP, LORD (1947). The preservation of bacteria by drying. *J. gen. Microbiol.*, **1,** 251.

WICKERHAM, L. J. & ANDREASEN, A. A. (1942). The lyophil process: its use in the preservation of yeasts. *Wallerstein Lab. Comm.*, **5,** 165.

The Disinfection of Heat Sensitive Surgical Instruments

V. G. ALDER AND J. P. MITCHELL

United Bristol Hospitals, Bristol BS2 8HW, England

The most effective method of sterilizing any instrument is by heat but endoscopic instruments cannot be autoclaved or exposed to hot air. Boiling water subjects the telescopes to violent and harmful agitation. Pasteurization in water at 80° is one recommended procedure (Francis, 1959) but the instrument is wet and cannot be stored in a disinfected condition. Organisms embedded in organic matter may be protected from the action of chemical disinfectants, moreover air bubbles inside the sheath of the endoscope may prevent the disinfecting fluid from reaching the entire lumen of the instrument (Miller *et al.*, 1960). The popular method of immersing the instruments in 0·5% Hibitane (chlorhexidine digluconate) in 70% ethanol is a rapid disinfection procedure but suffers from the above mentioned limitations and is also damaging to lens mountings. Ethylene oxide gas is not practical for routine purposes in the operating theatre, mainly due to the slowness of the method and the fact that the instruments must be scrupulously clean before exposure to the gas. Following urological surgery, endoscopes (Figs 3 and 4) may become contaminated with organisms embedded in organic material inside the crevices of the metal junctions of the instrument, in the interfacial surfaces of metal taps and inside the sheath. The instruments are difficult to clean thoroughly and too vigorous cleaning methods may damage lens mountings.

Ideally any method of sterilization for endoscopic instruments must be reliably bactericidal and avoid chemical and heat trauma to the instrument. The method should have a rapid time cycle and be standard for all parts of the instrument. Preferably it should allow prepacking of the instrument and safe storage in a dry sterile state for transport and ready availability. Disinfection of endoscopic instruments by the low temperature steam method, described below, combines all these advantages and is economical, easy to operate and control.

The use of low temperature steam at 80° was developed for the purpose of disinfecting heat sensitive hospital equipment (Alder and Gillespie, 1961).

At first the method was used for disinfecting woollen blankets (Alder and Leitch, 1963) but, because of the speed and efficiency of the process, the method was successfully employed for disinfecting the endoscopic instruments and electrical attachments which are used in urological surgery (Mitchell, 1970). The steam process destroys all vegetative cells within 3 min exposure at 80° and the addition of small quantities of formaldehyde to the steam renders the process sporicidal within 2 h of exposure to the gases (Alder, Brown and Gillespie, 1966; Gibson, Johnston and Turkington, 1968). The gases penetrate several layers of fabric and inside lengths of fine bore tubing.

The problem arose of how to test a complicated piece of equipment such as an endoscope for sterility, and this paper describes the bacteriological test methods which were developed in order to evaluate the efficiency of this new method of disinfection. The paper also describes the apparatus and the technique of operating the disinfector.

Materials and Methods

Principle of the method

The principle of the method is based on the fact that the temperature of steam depends upon its pressure, so that if steam or a steam formaldehyde mixture is admitted to a previously evacuated chamber the temperature inside the vessel can be accurately maintained by the simple process of controlling the pressure. Thus with a steam and formaldehyde mixture the optimum temperature required to obtain the maximum sporicidal effect with minimum damage to the instruments was found to be 80°, which is given (Fig. 1) by a pressure of 355 mm Hg (approx. half an atmosphere).

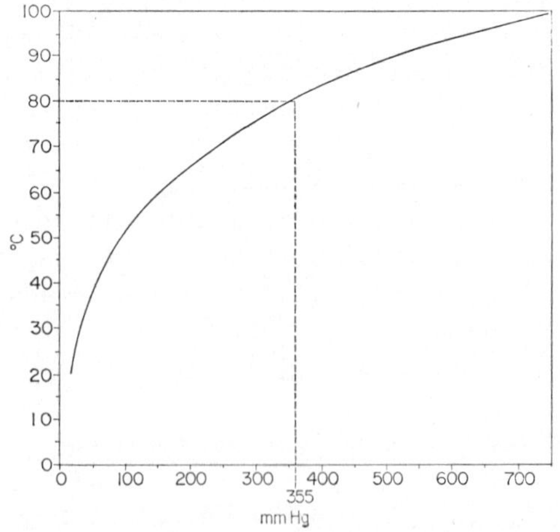

FIG. 1. Temperature of saturated steam at sub-atmospheric pressure.

The apparatus

A diagram of the apparatus originally developed in Bristol is shown in Fig. 2. The apparatus, connected to a mains steam supply, consisted of a horizontal rectangular jacketed autoclave fitted with a valve for controlling the steam pressure at 355 mm Hg, a vapourizer for generating formaldehyde from formalin, and a high vacuum pump and condenser. The jacket

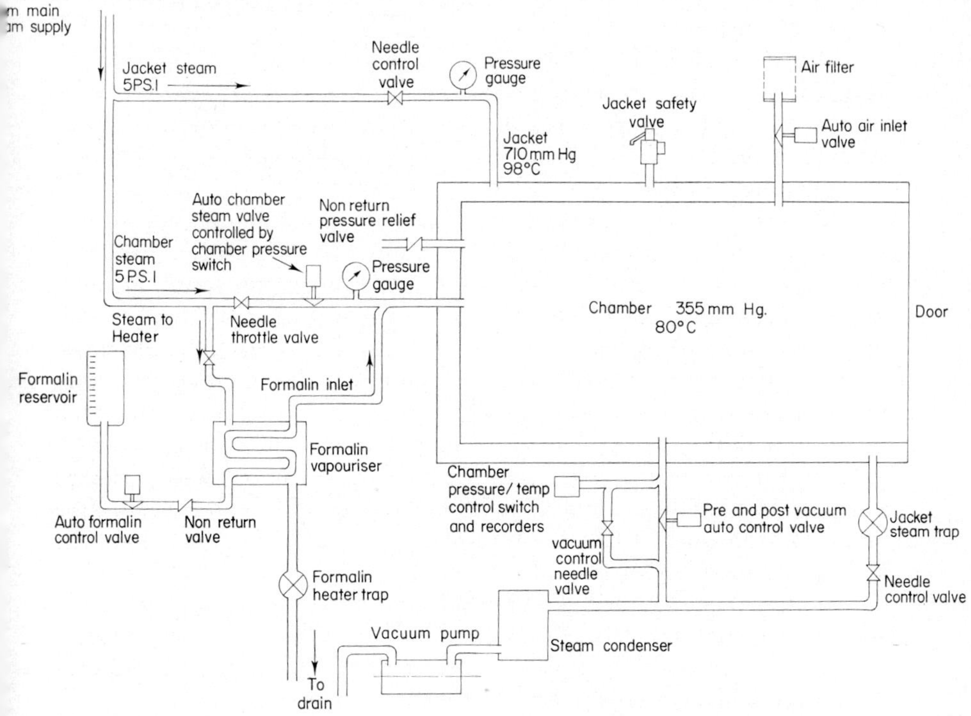

FIG. 2. Diagram of a sub-atmospheric steam/formaldehyde disinfector.

temperature was maintained at 98°, the chamber was maintained at 80°. There was no risk of damage to packaged instruments by superheat from the jacket wall during disinfection.

The endoscopes

The instruments used were of British, German and American make and were tested over a period of three years routine hospital use. Fibre-lit telescopes and cables were tested occasionally.

The method of operating the disinfector

Disinfection by steam and formaldehyde at 80°

1. The chamber (Fig. 2) was loaded and the door closed. The vacuum pump is started and remains working for the duration of the process, to assist removal of condensate from the chamber, through the needle control valves.

2. Air was evacuated from the chamber to a final positive pressure of 12 mm Hg or less.

3. Formalin (37% w/v H.CHO), 2 ml/ft^3 of autoclave space, was admitted to the chamber through the vaporizer, and the chamber was then re-evacuated. This process was repeated twice, which removed traces of air and ensured a thorough penetration of formaldehyde vapour into the packaged instruments.

4. Formalin, 8 ml/ft^3, of autoclave space, was then admitted through the vaporizer to the chamber, followed by steam until a pressure of 355 mm Hg (approx. half an atmosphere) was reached inside the vessel, giving a temperature reading of 80°. This pressure was maintained for 2 h and any fall in pressure due to condensation was automatically corrected by the steam valve admitting the required quantity of steam. It was important to admit the formaldehyde to the chamber before the steam to ensure a thorough penetration of the gas, after which no more formaldehyde was necessary.

5. At the end of the 2 h exposure period the chamber was evacuated twice by the vacuum pump to a minimum pressure of 15 mm Hg admitting steam to a pressure of half an atmosphere between each evacuation. The vacuum pump was switched off.

6. Filtered air was then admitted to the chamber to atmospheric pressure and the dry materials inside were removed. A faint smell of formaldehyde was sometimes present when the door was opened due to the absorption of formaldehyde round the autoclave door washers.

This process was sporicidal.

Disinfection by steam only at 80°

1. The autoclave chamber was loaded and the door closed. The vacuum pump was started and remained working for the duration of the process, to assist the removal of condensate from the chamber through the needle control valves.

2. Air was evacuated from the chamber to a final positive pressure of 12 mm Hg or less.

3. Steam was admitted to the chamber until a pressure of 355 mm Hg was reached and was maintained for 5 min.

4. At the end of the 5 min period the chamber was evacuated by the

high vacuum pump to a pressure of 15 mm Hg. The vacuum pump was switched off.

5. Filtered air was then admitted to the chamber to atmospheric pressure, and the dry materials inside were removed.

This process killed all vegetative cells but not spores.

Test pieces (simulators)

In order to evaluate the efficiency of a disinfection method for endoscopes, it was necessary to test for bacterial survival in conditions likely to be found on the instrument. An attempt to solve this problem was made by constructing test pieces, simulating the most protected areas of endoscopes but which could be submitted fairly easily to bacteriological examination.

One test piece consisted of a screw-capped 1 oz (28 ml capacity) glass container, with 17·5 cm of 1 mm bore tubing soldered into the lid (Fig. 3). The steam and formaldehyde had to pass along the tubing to enter the chamber, within which were placed cultures of bacteria in small plugged glass test tubes. In further tests to simulate the conditions of interfacial contamination inside taps, the bacteria were smeared on the threads of a nut and bolt which was assembled after drying.

Both kinds of test pieces were sealed inside paper bags and placed in cardboard cartons and exposed to steam or steam and formaldehyde. Afterwards the bacterial cultures were tested for survivors. It was concluded that if the organisms were killed inside the test piece one could be reasonably certain that they would be killed inside an endoscope (Mitchell and Alder, 1970).

The test organisms

The routine test organisms used were cultures of *Streptococcus faecalis* and *Bacillus stearothermophilus* spores. The organisms were suspended in 10 ml of a medium consisting of equal parts of normal Horse Serum and 0·85% (w/v) NaCl in water to give counts of *c.* 10^5 to 10^6/ml.

About 0·1 ml of the suspensions was dried on 25 × 6 mm paper strips over $CaCl_2$ in a desiccator and placed in 53 × 9 mm plugged glass tubes.

Tests for surviving organisms

B. stearothermophilus spore cultures were used for testing the sporicidal efficiency of steam/formaldehyde at 80°, survivors were detected by incubation in Tryptone-Dextrose Broth at 56° for 5 days. *Streptococcus faecalis* cultures were used for testing the bactericidal efficiency of steam at

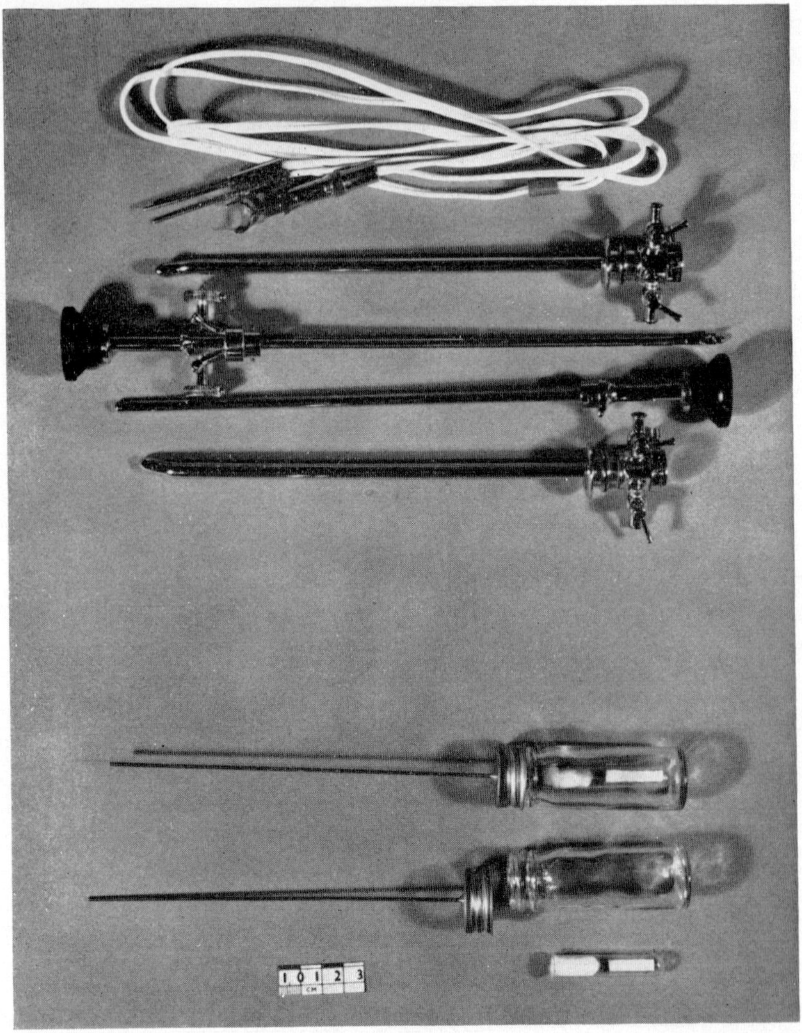

FIG. 3. Endoscopes and simulator test pieces. (One test piece dismantled with culture tube removed.)

80°; survivors were detected by incubation in Nutrient Broth at 37° for 5 days.

Results

Non-sporing organisms

In numerous experiments cultures of *Str. faecalis*, dried from serum and saline suspensions, were killed within 5 min by steam only at 80°

even when sealed inside the threads of an assembled nut and bolt. Heat from the steam is transmitted through the metal and kills the organism. In other experiments cultures of *Staphylococcus aureus* and *Escherichia coli*, similarly treated, were also killed within 5 min exposure.

Sporing organisms

Routine monitoring of the process showed that steam/formaldehyde mixtures at 80° killed cultures of *B. stearothermophilus*, dried from serum and saline suspensions, within 2 h. The process failed to kill the spores when they were entrapped between the threads of an assembled nut and bolt. Heat sterilization methods using higher temperatures would be required to kill sporing organisms on interfacial metal surfaces.

The endoscopes

All the instruments were undamaged by the process after three years of routine hospital use and no corrosion of electrical attachments has been observed.

Residual formaldehyde

With other methods of sterilization by formaldehyde for electrical equipment, objections have been raised because the breakdown products of formaldehyde to methanol and formic acid eroded electrical connections (Opfell and Miller, 1965). In contrast, chemical tests showed that after steam/formaldehyde disinfection at 80° only minute quantities of formalin, in the order of 4 to 8 ppm, remained on rubber and polythene (Gibson, Johnston and Turkington, 1968) and on the metal endoscopic instruments in routine use and the electrical leads and switches there was no evidence of corrosion.

Packaging

The endoscopes, complete with electric light bulbs, leads and switches, were prepackaged in layers of plastic foam and enclosed in a cardboard box (Fig. 4) before disinfection. In this way, after disinfection the instruments were stored in a sterile state for several days before subsequent usage, thus conforming to modern central sterile supply practices.

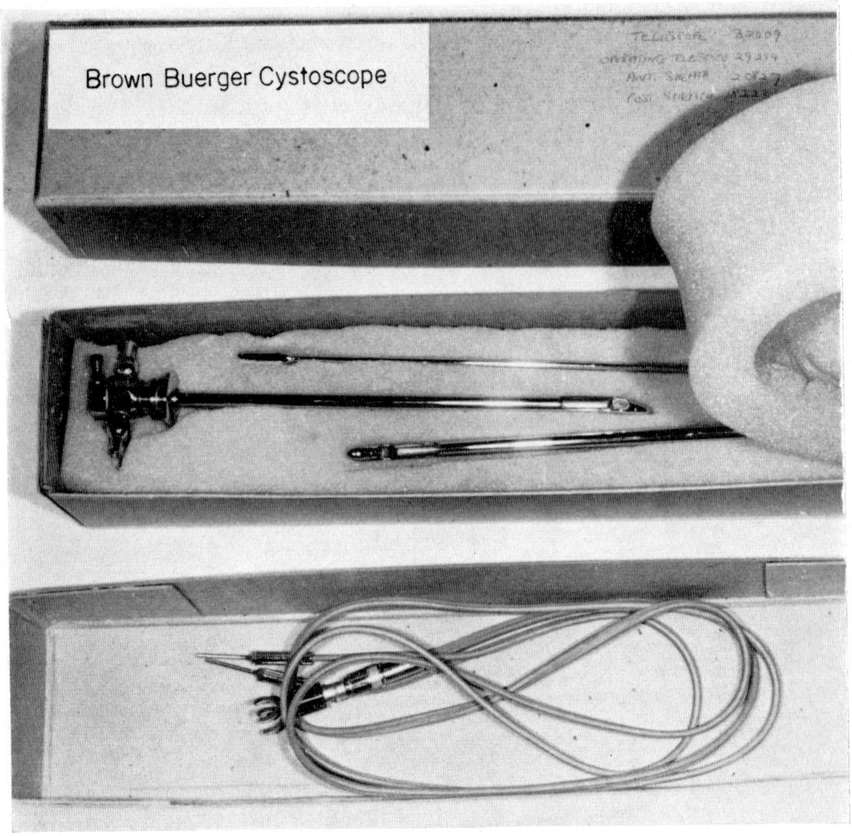

FIG. 4. Packaging of an endoscope in plastic foam and cardboard carton.

Conclusions

The low temperature steam and formaldehyde process of disinfecting heat sensitive endoscopic instruments is bacteriologically efficient, economical and easy to operate. The process could possibly be used for sterilizing other heat sensitive equipment.

Bacterial spores can survive the process if they are enclosed by interfacial metal surfaces (e.g. the threads of assembled nuts and bolts and the interfacial surfaces of close fitting taps) but sufficient heat penetrates the metal to kill vegetative cells in 3 min.

Since sporing organisms rarely infect the urinary tract and the risk of transmitting sporing organisms by an endoscope is negligible, steam only at 80° may be used for disinfecting these instruments (Anon, 1967). This process has been in use at the United Bristol Hospitals since 1965, catering for 5 urological lists per week. The disinfection cycle is rapid and need take no longer than 15 min and is carried out between cases in an automatically operated disinfector.

At the end of the operating list, the sporicidal and longer steam/formaldehyde process can be employed, and the instruments stored ready for immediate use at the next operating list.

Acknowledgements

We gratefully acknowledge the help and advice received from Professor W. A. Gillespie, Professor of Clinical Bacteriology, University of Bristol; Dr. J. C. Kelsey, Director, Disinfection Reference Laboratory, Central Public Health Laboratories, London N.W.9.; Mr. A. Bishop, Department of Health and Social Security, London W.C.2. and Mr. M. I. Lees, Group Engineer, United Bristol Hospitals.

References

Alder, V. G. & Gillespie, W. A. (1961). Disinfection of woollen blankets in steam at sub-atmospheric pressure. *J. Clin. Path.*, **14**, 515.

Alder, V. G. & Leitch, C. W. (1963). Disinfection of blankets by low temperature steam. *Hospital, Lond.*, **59**, 662.

Alder, V. G., Brown, Anne M. & Gillespie, W. A. (1966). Disinfection of heat sensitive material by low temperature steam and formaldehyde. *J. Clin. Path.*, **19**, 83.

Anon. (1967). Notes on Sterilisation No. 1. *Sterilisation and Disinfection of Cystoscopes,* The Association of Clinical Pathologists. London: B.M.A. House.

Francis, A. E. (1959). Disinfection of cystoscopes by pasteurisation. *Proc. Roy. Soc. Med.* **52,** 998.

Gibson, G. L., Johnston, H. P. & Turkington, V. E. (1968). Residual formaldehyde after low-temperature steam and formaldehyde sterilisation. *J. clin. Path.*, **21**, 771.

Miller, A., Gillespie, W. A., Linton, K. B., Slade, N. & Mitchell, J. P. (1960). Prevention of urinary infection after prostatectomy. *Lancet,* **2,** 886.

Mitchell, J. P. (1970). Transurethral resection. *Br. med. J.*, **11**, 241.

Mitchell, J. P. & Alder, V. G. (1970). Recent developments on the use of subatmospheric steam and formaldehyde gas at 80°C for the disinfection of cystoscopes. *Br. Hosp. J. Social Serv. Rev.*, **LXXX,** 1944.

Opfell, J. B. & Miller, C. B. (1965). Cold sterilisation technique. *Advances in Applied Microbiology,* Vol. 7. New York and London: Academic Press.

Author Index

Numbers in italics are pages on which references are listed at the end of the paper.

Subject Index